Autonomous Airborne Wireless Networks

Autonomous Airborne Wireless Networks

Edited by
Muhammad Ali Imran, Oluwakayode Onireti,
Shuja Ansari, and Qammer H. Abbasi

University of Glasgow, UK

IEEE PRESS
WILEY

The right of Muhammad Ali Imran, Oluwakayode Onireti, Shuja Ansari, and Qammer H. Abbasi to be identified as the editors of this work has been asserted in accordance with law.

Registered Offices
John Wiley & Sons, Inc., 111 River Street, Hoboken, NJ 07030, USA
John Wiley & Sons Ltd, The Atrium, Southern Gate, Chichester, West Sussex, PO19 8SQ, UK

Editorial Office
The Atrium, Southern Gate, Chichester, West Sussex, PO19 8SQ, UK

For details of our global editorial offices, customer services, and more information about Wiley products visit us at www.wiley.com.

Wiley also publishes its books in a variety of electronic formats and by print-on-demand. Some content that appears in standard print versions of this book may not be available in other formats.

Library of Congress Cataloging-in-Publication Data Applied for:
ISBN: 9781119751687

Cover Design: Wiley
Cover Images: © Shine Nucha/Shutterstock, © Solveig Been/Shutterstock

Set in 9.5/12.5pt STIXTwoText by Straive, Chennai, India
Printed and bound by CPI Group (UK) Ltd, Croydon, CR0 4YY

C9781119751687_220721

Contents

**9   Performance Analysis of UAV-Enabled Disaster
           Recovery Networks**   *157*
           *Rabeea Basir, Saad Qaisar, Mudassar Ali, Naveed Ahmad Chughtai,
           Muhammad Ali Imran, and Anas Hashmi*

Editor Biographies

Muhammad Ali Imran is Dean at the University of Glasgow, UESTC, Professor of Communication Systems, and Head of Communications Sensing and Imaging group in the James Watt School of Engineering at the University of Glasgow, UK.

Oluwakayode Onireti is a lecturer at the James Watt School of Engineering, University of Glasgow, UK.

Shuja Ansari is currently a research associate at Communications Sensing and Imaging group in the James Watt School of Engineering at the University of Glasgow, UK, and Wave-1 Urban 5G use case implementation lead at Glasgow 5G Testbed funded by the Scotland 5G Center.

Qammer H. Abbasi is a senior lecturer (Associate Professor), Program Director for Dual PhD degree, and Deputy Head of Communications Sensing and Imaging group in the James Watt School of Engineering at the University of Glasgow, UK.

List of Contributors

Qammer H. Abbasi
James Watt School of Engineering
University of Glasgow
Glasgow
UK

Hisham Abuella
School of Electrical and Computer
Engineering, Oklahoma State
University
Stillwater, OK
USA

Rigoberto Acosta-González
Department of Electronics and
Telecommunications, Universidad
Central "Marta Abreu" de Las Villas
Santa Clara
Cuba

Muhammad W. Akhtar
School of Electrical Engineering
and Computer Science (SEECS)
National University of Sciences and
Technology (NUST)
Islamabad
Pakistan

Gotta Alberto
Institute of Information Science
and Technologies (ISTI) and
Institute of Science and
Technologies for Energy and
Sustainable Mobility, National
Research Council (CNR)
Pisa
Italy

Mudassar Ali
Department of Telecommunication
Engineering, UET
Taxila
Pakistan

Imran S. Ansari
James Watt School of Engineering
University of Glasgow
Glasgow
UK

Rafay I. Ansari
Department of Computer and
Information Science
Northumbria University
Newcastle upon Tyne
UK

Shuja S. Ansari
James Watt School of Engineering
University of Glasgow
Glasgow
UK

Muhammad R. Asghar
School of Computer Science
The University of Auckland
Auckland
New Zealand

Muhammad Awais
School of Computing and
Communications
Lancaster University
Lancaster
UK

Elizabeth Basha
Electrical and Computer
Engineering Department
University of the Pacific
Stockton, CA
USA

Rabeea Basir
School of Electrical Engineering &
Computer Science (SEECS)
National University of Sciences and
Technology
Islamabad
Pakistan

and

James Watt School of Engineering
University of Glasgow
Glasgow
UK

Charles F. Bunting
School of Electrical and Computer
Engineering, Oklahoma State
University
Stillwater, OK
USA

Yunfei Chen
School of Engineering
University of Warwick
Coventry
UK

Naveed A. Chughtai
Military College of Signals
National University of Sciences and
Technology
Rawalpindi
Pakistan

Jacob N. Dixon
IBM
Rochester, MN
USA

Sabit Ekin
School of Electrical and Computer
Engineering, Oklahoma State
University
Stillwater, OK
USA

Syed A. Hassan
School of Electrical Engineering
and Computer Science (SEECS)
National University of Sciences and
Technology (NUST)
Islamabad
Pakistan

Muhammad A. Imran
James Watt School of Engineering
University of Glasgow
Glasgow
UK

Jamey D. Jacob
School of Mechanical and
Aerospace Engineering, Oklahoma
State University
Stillwater, OK
USA

Dushantha Nalin K. Jayakody
Department of Information
Technology, School of Computer
Science and Robotics, National
Research Tomsk Polytechnic
University
Tomsk
Russian Federation

and

Centre for Telecommunication
Research, School of Engineering
Sri Lanka Technological Campus
Padukka
Sri Lanka

Amit Kachroo
School of Electrical and Computer
Engineering, Oklahoma State
University
Stillwater, OK
USA

Aziz Khuwaja
School of Engineering, Electrical
and Electronic Engineering Stream
University of Warwick
Coventry
UK

Paulo V. Klaine
Electronics and Nanoscale
Engineering Department
University of Glasgow
Glasgow
UK

Hassan Malik
Department of Computer Science
Edge Hill University
Ormskirk
UK

Bacco Manlio
Institute of Information Science
and Technologies (ISTI) and
Institute of Science and
Technologies for Energy and
Sustainable Mobility, National
Research Council (CNR)
Pisa
Italy

Ruggeri Massimiliano
National Research Council (CNR)
Institute of Science and
Technologies for Energy and
Sustainable Mobility
Ferrara
Italy

Lina Mohjazi
James Watt School of Engineering
University of Glasgow
Glasgow
UK

Samuel Montejo-Sánchez
Programa Institucional de Fomento
a la I+D+i, Universidad
Tecnológica Metropolitana
Santiago
Chile

Hieu V. Nguyen
The University of
Danang – Advanced Institute of
Science and Technology
Da Nang
Vietnam

Qiang Ni
School of Computing and
Communications
Lancaster University
Lancaster
UK

Phu X. Nguyen
Department of Computer
Fundamentals, FPT University
Ho Chi Minh City
Vietnam

Van-Dinh Nguyen
Interdisciplinary Centre for
Security, Reliability and Trust
(SnT), University of Luxembourg
Luxembourg

Oluwakayode Onireti
James Watt School of Engineering
University of Glasgow
Glasgow
UK

and

Department of Electrical
Engineering, Sukkur IBA
University
Sukkur
Pakistan

Barsocchi Paolo
Institute of Information Science
and Technologies (ISTI) and
Institute of Science and
Technologies for Energy and
Sustainable Mobility, National
Research Council (CNR)
Pisa
Italy

Haris Pervaiz
School of Computing and
Communications
Lancaster University
Lancaster
UK

Olaoluwa Popoola
James Watt School of Engineering
University of Glasgow
Glasgow
UK

Tharindu D. Ponnimbaduge Perera
Department of Information
Technology, School of Computer
Science and Robotics, National
Research Tomsk Polytechnic
University
Tomsk
Russian Federation

Adithya Popuri
School of Electrical and Computer
Engineering, Oklahoma State
University
Stillwater, OK
USA

Saad Qaisar
School of Electrical Engineering &
Computer Science (SEECS)
National University of Sciences and
Technology
Islamabad
Pakistan

and

Department of Electrical and
Electronic Engineering
University of Jeddah
Jeddah
Saudi Arabia

Marwa Qaraqe
Division of Information and
Computing Technology, College of
Science and Engineering, Hamad
Bin Khalifa University (HBKU)
Doha
Qatar

Navuday Sharma
Test Software Development
Ericsson Eesti AS
Tallinn
Estonia

Richard D. Souza
Department of Electrical and
Electronics Engineering, Federal
University of Santa Catarina
Florianóplis
Brazil

Muhammad K. Shehzad
School of Electrical Engineering
and Computer Science (SEECS)
National University of Sciences and
Technology (NUST)
Islamabad
Pakistan

Oh-Soon Shin
School of Electronic Engineering
Soongsil University
Seoul
South Korea

Sean Thalken
Electrical and Computer
Engineering Department
University of the Pacific
Stockton, CA
USA

Jason To-Tran
Electrical and Computer
Engineering Department
University of the Pacific
Stockton, CA
USA

Christopher Uramoto
Electrical and Computer
Engineering Department
University of the Pacific
Stockton, CA
USA

Muhammad Usman
Division of Information and
Computing Technology, College of
Science and Engineering, Hamad
Bin Khalifa University (HBKU)
Doha
Qatar

Surbhi Vishwakarma
School of Electrical and Computer
Engineering, Oklahoma State
University
Stillwater, OK
USA

Davis Young
Electrical and Computer
Engineering Department
University of the Pacific
Stockton, CA
USA

Lei Zhang
Electronics and Nanoscale
Engineering Department
University of Glasgow
Glasgow
UK

1

Introduction

Muhammad A. Imran, Oluwakayode Onireti, Shuja S. Ansari, and Qammer H. Abbasi

James Watt School of Engineering, University of Glasgow, Glasgow, UK

Airborne networks (ANs) are now playing an increasingly crucial role in military, civilian, and public applications such as surveillance and monitoring, military, and rescue operations. More recently, airborne networks have also become a topic of interest in the industrial and research community of wireless communication. The 3rd Generation Partnership Project (3GPP) standardization has a study item devoted to facilitating the seamless integration of airborne wireless networks into future cellular networks. Airborne wireless networks enabled by unmanned aerial vehicles (UAVs) can provide cost-effective and reliable wireless communications to support various use cases in future networks. Compared with high-altitude platforms or conventional terrestrial communications, the provision of on-demand communication systems with UAVs has faster deployment time and more flexibility in terms of reconfiguration. Further, UAV-enabled propagation can also offer better communication channels due to the existence of the line-of-sight (LoS) links, which are of short range.

Despite the several benefits of airborne wireless networks, they suffer from some realistic constraints such as being energy constrained because of the limited battery power, safety concerns, and the strict flight zone. Hence, developing new signal processing, communication, and optimization framework for autonomous airborne wireless networks is essential. Such networks can offer high data rates and assist the traditional terrestrial networks to provide real-time and ultrareliable sensing applications for the beyond-5G networks. Achieving this gain requires the correct characterization of the propagation channel while considering the high

Autonomous Airborne Wireless Networks, First Edition.
Edited by Muhammad Ali Imran, Oluwakayode Onireti, Shuja Ansari, and Qammer H. Abbasi.

mobility dynamics. Accurate channel modeling is imperative to fulfill the ever-increasing requirements of the end user to transfer data at higher rates. The air-to-ground (AG) and the air-to-air (AA) channel propagation models for the airborne wireless network channel can be characterized by using measurement and empirical studies. Further, the key performance indicators (KPIs) of airborne wireless networks such as flight time, trajectory, data rate, energy efficiency, and latency need to be optimized for the different use cases.

This book explores recent advances in the theory and practice of airborne wireless networks for the next generation of wireless networks to support various applications, including emergency communications, coverage and capacity expansion, Internet of things (IoT), information dissemination, future healthcare, pop-up networks, etc. The book focuses on channel characteristics and modeling, networking architectures, self-organized airborne networks, self-organized backhaul, artificial-intelligence-enabled trajectory optimization, and application in sectors such as agriculture, underwater communications, and emergency networks. The book further highlights the main considerations during the design of the autonomous airborne networks and exploits new opportunities due to the recent advancement in wireless communication systems.

This book for the first time evaluates the advances in the current state of the art and it provides readers with insights on how airborne wireless networks can seamlessly support various applications expected in future networks. More specifically, the book shows the readers how the integration of self-organized networks and artificial intelligence can support the various use cases of airborne wireless networks.

UAVs provide a suitable aerial platform for various wireless network applications that require reliable and ubiquitous communication. The channel model plays a crucial role in the wireless communications system and thus Chapter 2 focuses on the channel model for UAV networks. The authors first provide an overview of UAV networks in terms of their classification and how they can be used to enable future wireless communication systems. Accurate channel modeling is imperative to fulfill the ever-increasing requirements of the end user to transfer data at higher rates. Hence, the authors discuss channel modeling in UAV communications while focusing on the salient feature of the AG and AA propagation channels. Finally, the chapter concludes by discussing some of the key research challenges for the practical deployment of UAVs as airborne wireless nodes.

In Chapter 3, the authors describe the fundamental properties of the ultrawide band (UWB) channel and present one of the first experimental off-body studies between a human subject and an UAV at 7.5 GHz of bandwidth. In the study presented in this chapter, the transmitter antenna was placed on a UAV while the receiver antenna was patched on a human subject at different body locations during the campaign. The chapter presents the measurement setting, detailing the measurement campaign that was conducted in an indoor and an outdoor environment with LoS and non-line-of-sight (NLoS) cases. Furthermore, the chapter presents the UWB-unmanned aerial vehicle-to-wearables (UAV2W) channel characterization. Finally, the chapter presents the statistical analysis to determine the distribution that best characterizes the fading channels between different body locations and the UAV.

Chapter 4 describes the use of a Q-learning algorithm, which is based on a cooperative multiagent approach, to intelligently find the optimal position of a set of drones. The algorithm presented in the chapter is designed with the objective to minimize the number of users in an outage in the network. Hence, the algorithm determines the optimal distribution of frequencies and whether it should shut down a set of drones. The chapter also proposes and compares four different strategies for the Q-learning algorithm with different action selection policies, whose algorithms differ in terms of design complexity, ability to vary the number of drones in operation, and convergence time. The chapter presents numerical results that show the relationship between the density of users in the region of interest and the number of frequencies in operation.

In Chapter 5, the authors describe a self-energized UAV-assisted caching relaying scheme. In this scheme, the UAV's communication capabilities are powered solely by the power-splitting simultaneous wireless information and power transfer (PS-SWIPT) energy-harvesting (EH) technique, and it employs decode and forward (DF) relaying protocol to assist the information transmission to users from the source node. The authors present the transmission block diagram to accommodate communication processes within the system. Afterward, the authors address the problem of identifying optimal time and energy resources for the communication system and the optimal UAV's trajectory while adhering to the quality of service (QoS) requirements of the communication network. Finally, numerical simulation results to identify the impacts of the system parameters on the information rate at the user equipment are presented.

Chapter 6 focuses on the case study of millimeter-wave (mmWave) and terahertz (THz) communication and technical challenges for applying mmWave and THz frequency band for communication with UAVs. The chapter starts by presenting the potential of mmWave and THz bands for communications. This is followed by an overview of the technical challenges for implementing mmWave and THz band for UAV communications.

The chapter then presents a theoretical analysis that focuses on the placement of UAVs. Besides, the chapter investigates the performance of UAV-enabled hybrid heterogeneous network (HetNet) by considering stringent communication-related constraints such as the system bandwidth, data rate, signal-to-noise ratio (SNR), etc. The association of terrestrial small-cell base stations (SCBs) with UAVs is addressed such that the sum rate of the overall system is maximized. Finally, numerical results are included to show the favorable performance of the UAV-assisted wireless network.

In Chapter 7, the authors discuss a method that uses a cooperative UAV as a friendly jammer to enhance the security performance of cognitive radio networks. The chapter starts by presenting the system model for the UAV-enabled cooperative jamming in a cognitive radio system. Then the optimization problem is formulated. The resource allocation in the network must jointly optimize the transmission power and UAV's trajectory to maximize the secrecy rate while satisfying a given interference threshold at the primary receiver (PR). With the original problem non-convex, the authors first transform the original problem into a more tractable form and then present a successive convex approximation-based algorithm for its solutions. Finally, numerical results are included to show a significant improvement in the security performance of the UAV-enabled cognitive radio networks.

Chapter 8 explores the possibility of using intelligent reflecting surfaces (IRS) in airborne networks for the localization of users and base stations. Positioning is an important aspect in the present and future wireless networks, where it augments the network operations and assists in multiple localization-based applications. The chapter starts by presenting the related works and the underlying opportunities around IRS- and UAV-based base stations. The authors then discuss the integration of IRS in ANs and the potential use cases. Afterward, the chapter presents an IRS-based localization model for ANs along with some mathematical modeling. Finally, some future research challenges that present research opportunities are included.

Chapter 9 describes the application of UAVs for disaster recovery networks. The chapter starts by providing an overview of the UAV networks including the description of the UAV architectures, namely, single-UAV systems, multi-UAV systems, cooperative multi-UAV systems, and multilayer UAV networks. The authors then discuss the most prominent applications of UAVs and the different system requirements of the UAV system. Afterward, the chapter discusses the design consideration of UAV networks in the context of disaster recovery networks. New technologies and infrastructure trends for UAV disaster networks namely, network function virtualization (NFV), software-defined networks (SDN), cloud computing, and millimeter-wave networks are also discussed in the chapter. Further, the authors discuss the enhancement in technologies such as artificial intelligence, machine learning, optimization theory, and game theory as they impact the overall performance of the UAV-enabled disaster recovery networks. Finally, the chapter presents the research trends and some insight into the future.

In Chapter 10, the authors discuss the importance of UAVs in monitoring COVID-19 restrictions of social distancing, public gatherings, and physical contacts in a smart city environment. The chapter starts with a review of recent literature addressing the impact of COVID-19 in the current scenario and strategies to find potential solutions with existing communication and computing technologies. Afterward, the authors present two use case scenarios of UAVs namely, UAVs as aerial base stations (ABS) and UAVs as Relays, while including the simulation setups with ray tracing for both scenarios. The chapter then presents the derivation of the optimal number of ABSs to cover a geographical region, given the constraint on ABS transmission power, the altitude of hovering, and including the path loss and channel fading effects from ray-tracing simulations. The authors then describe the 5G air interface when using the UAVs as relays. Finally, simulation results on the received power by the ground users and the throughput coverage area are presented.

In Chapter 11, the authors present and discuss both the research initiatives and the scientific literature on IoT-based smart farming (SF), especially the use of UAVs in SF. The authors start by presenting an analysis of how UAVs are used in SF and the application scenarios. This is then followed by a detailed review of the scientific work in the literature highlighting the role of unmanned vehicles. The chapter then presents both the requirements and solutions for networking and a brief comparison of the existing protocol supporting IoT scenarios in agricultural settings. Finally, the chapter discusses the potential future role of the joint use of mobile edge computing (MEC) and the 5G network, presenting network architecture to connect smart farms through UAVs and satellites.

Wetlands monitoring requires accurate topographic and bathymetric maps, and this can be achieved using UAVs that can create maps regularly, with minimum cost and reduced environmental impact. Chapter 12 introduces a set of systems needed to create this automation. The chapter starts by discussing the automated image labeling system. Next, the authors present an online classification system for differentiating land and water. The authors then present offline bathymetric map creation using aerial robots. Since the offline approach does not take full advantage of the adaptability that the UAV provides, the authors present the online bathymetric mapping. Finally, the chapter presents results and analysis to show the best combination of the online bathymetric mapping.

Integration of terrestrial and satellite networks has been proposed for leveraging the combined benefits of both complementary technologies. Moreover, with the quest of exploring deep space and connecting solar system planets with the Earth, the traditional satellite network has gone beyond the geosynchronous equatorial orbit (GEO) wherein Interplanetary Internet will play a key role. Chapter 13 presents a short review of the inter-satellite and deep space network (ISDSN). This chapter discusses the classification of the ISDSN into different tiers while highlighting the communication and networking paradigms. Further, the chapter also discusses the security requirements, challenges, and threats in each tier. The potential solutions to the identified challenges at the different tiers of the ISDSN are also described. Finally, the chapter concludes by highlighting the crucial role of the ISDSN in future cellular networks.

2

Channel Model for Airborne Networks

Aziz A. Khuwaja[1,2] and Yunfei Chen[1]

[1] School of Engineering, Electrical and Electronic Engineering Stream, University of Warwick, Coventry, UK
[2] Department of Electrical Engineering, Sukkur IBA University, Sukkur, Pakistan

2.1 Introduction

The use of unmanned aerial vehicles (UAVs) is desirable due to their high maneuverability, ease of operability, and affordable prices in various civilian applications, such as disaster relief, aerial photography, remote surveillance, and continuous telemetry. One of the promising application of UAVs is enabling the wireless communication network in cases of natural calamity and in hot spot areas during peak demand where the resources of the existing communication network have been depleted [1]. Qualcomm has already initiated field trials for the execution of fifth generation (5G) cellular applications [2]. Google and Facebook are also exploiting the use of UAVs to provide Internet access to far-flung destinations [3].

The selection of an appropriate type of UAV is essential to meet the desired quality of service (QoS) depending on applications and goals in different environments. In fact, for any specific wireless networking application, the UAV altitude and its capabilities must be taken into account. UAVs can be categorized, based on their altitude, into low-altitude platforms (LAPs) and high-altitude platforms (HAPs). Furthermore, based on their structure, UAVs can be categorized as fixed-wing and rotary-wing UAVs. In comparison with rotary wings, fixed-wing UAVs move in the forward direction to remain aloft, whereas rotary-wing UAVs are desired for applications that require UAVs to be quasi-stationary over a given area. However, in both types, flight duration depends on their energy sources, weight, speed, and trajectory.

Autonomous Airborne Wireless Networks, First Edition.
Edited by Muhammad Ali Imran, Oluwakayode Onireti, Shuja Ansari, and Qammer H. Abbasi.
© 2021 John Wiley & Sons Ltd. Published 2021 by John Wiley & Sons Ltd.

The salient features of UAV-based communication network are the air-to-ground (AG) and air-to-air (AA) propagation channels. Accurate channel modeling is imperative to fulfill the ever-increasing requirements of end users to transfer data at higher rates. The available channel models for AG propagation are designed either for terrestrial communication or for aeronautical communications at higher altitudes. These models are not preferable for low-altitude UAV communication, which uses small size UAVs in different urban environments. On one hand, the AG channel exhibits higher probability of line-of-sight (LoS) propagation, which reduces the transmit power requirement and provides higher link reliability. In cases with non-line-of-sight (NLoS), shadowing and diffraction losses can be compensated with a large elevation angle between the UAV and the ground device. On the other hand, UAV mobility can incur significant temporal variations in both the AG and AA propagation due to the Doppler shift.

Small UAVs may experience airframe shadowing due to their flight path with sharper changes in pitch, yaw, and roll angle. In addition, distinct structural design and material of UAV body may contribute additional shadowing attenuation. This phenomenon has not yet been extensively studied in the literature.

Despite the number of promising UAV applications, one must address several technical challenges before the widespread applicability of UAVs. For example, while using UAV in aerial base station (BS) scenario, the important design considerations include radio resource management, flight time, optimal three-dimensional deployment of UAV, trajectory optimization, and performance analysis. Meanwhile, considering UAV in the aerial user equipment (UE) scenario, the main challenges include interference management, handover management, latency control, and three-dimensional localization. However, in both scenarios, channel modeling is an important design step in the implementation of UAV-based communication network. This chapter provides an overview of the use of UAV as aerial UEs and aerial BSs and discusses the technical challenges related to AG channel modeling, airframe shadowing, optimal deployment of UAVs, trajectory optimization, resource management, and energy efficiency.

2.2 UAV Classification

The need for an appropriate type of UAV depends on the specific mission, environmental conditions, and civil aviation regulations to attain certain

Table 2.1 Regulation for LAP deployment of UAVs in different countries.

Country	Maximum altitude (m)	Minimum distance to humans (m)	Minimum distance to airport (km)
US	122	—	8
UK	122	50	—
Chile	130	36	—
Australia	120	30	5.5
South Africa	46	50	10

altitude. In addition, for any particular UAV-enabled wireless networking application, several factors, such as the number of UAVs, their optimal deployment, and QoS requirement, must be taken in to account. The operational altitude of the UAV from the ground level can be categorized as LAP and HAP. UAVs in LAP can fly between the altitude ranges from tens of meters to a few kilometers [4]. However, civil aviation authorities of some countries have set the operational altitude of UAVs up to a few hundred meters to avoid airborne collision with commercial flights. For example, Table 2.1 lists the regulations of maximum allowable LAP deployment of UAVs in various countries without any specific permit [5]. HAPs, on the other hand, have altitudes above 17 km where UAVs are typically quasi-stationary [1, 4].

For time-sensitive applications such as emergency services, LAPs are more appropriate then HAPs due to their rapid deployment, quick mobility, and cost-effectiveness. Furthermore, LAPs can be used for collecting sensor data from the ground. In this case, LAPs can be readily replaced or recharged as needed. In contrast, HAPs are preferred due to their long endurance (days or months) operations and wider ground coverage [1]. However, operational cost of HAPs is high and their deployment time is significantly longer.

UAV can also be categorized based on their structure into rotary-wing and fixed-wing UAVs. Rotary-wing UAVs are powered by rotating blades, and based on the number of blades they are termed as either quadcopter with four blades, hexacopter with six blades, or octocopter with eight blades. On the other hand, fixed-wing UAVs include those that are driven by propellers with small size engine and have wings that are fixed. However, the flight time of UAVs relies on several key factors, such as type, weight, speed, energy sources (battery or engine), and trajectory of the UAV.

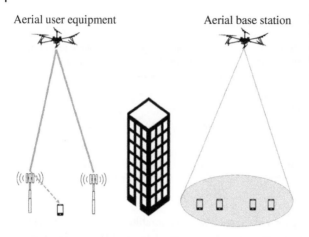

Figure 2.1 Aerial user equipment and aerial base station.

2.3 UAV-Enabled Wireless Communication

UAVs can operate as aerial UE as shown in Figure 2.1. For example, aerial surveillance can be a cost-effective solution to provide access to those terrains that may be difficult to reach by humans in land vehicles. In this case, UAVs equipped with camera and sensors are used to gather video recordings and live images of a specific target on the ground and data from the sensor. Thereafter, the UAV has to coordinate with the ground user via existing cellular infrastructure and transfer the collected information with certain reliability, throughput, and delay while achieving the QoS requirements. The first scenario in Figure 2.1 (left side) requires a better connectivity between the aerial UE and at least one of the BSs installed typically at the ground. However, a performance drop is expected in the presence of aerial BSs acting as interferers. Moreover, the coexistence between the aerial UE, terrestrial UE, and the cellular infrastructure has to be studied.

On the other hand, UAVs provide power efficiency and mobility to deploy as aerial BS in the future wireless networks. In this case, the mobility of UAV can dynamically provide additional on-demand capacity. This advantage of UAV-enabled network can be exploited by service providers for densification of network, temporary coverage of an area, or quick network deployment in an emergency scenario. Moreover, localization service precision can be improved due to the favorable propagation conditions between the UAV and the ground user. The second scenario in Figure 2.1 (right side) requires a better link between one of the multiple aerial BSs and all the terrestrial UEs. In comparison with fixed BSs, the aerial BSs are capable of adjusting their altitude to provide good LoS propagation. However, the key challenge in this scenario is the optimum placement of aerial BSs to maximize the ground coverage for higher achievable throughput.

2.4 Channel Modeling in UAV Communications

In wireless communications, the propagation channel is the free space between the transmitter and the receiver. It is obvious that the performance of wireless networks is influenced by the characteristics of the propagation channel. Therefore, knowledge of wireless channels is pertinent in designing UAV-enabled networks for future wireless communication. Furthermore, the characterization of radio channel and its modeling for UAV network architecture are crucial for the analysis of network performance.

Majority of the channel modeling efforts is devoted to the terrestrial radio channel with fixed infrastructure. However, these channel models may not be completely suitable for wireless communication using UAVs because of their mobility and small size. The AG channel between the UAV and the ground user implies higher link reliability and requires lower transmission power due to the higher probability of LoS propagation. In the case of NLoS, power variations are more severe because the ground-based side of the AG link is surrounded by obstacles that adversely affect the propagation. Figure 2.2 depicts the AG propagation channel and shows the distinction between LoS and NLoS components of the channel, with d_p being the propagation distance. Furthermore, temporal variations and the Doppler shift are caused by the UAV mobility. As a result, the arbitrary UAV mobility pattern and operational environment are challenges in modeling the AG

Figure 2.2
Air-to-ground propagation in UAV-assisted cellular network.

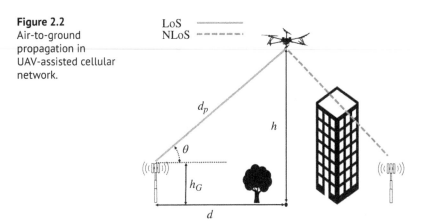

channel. Apart from the AG propagation channel, other factors such as airframe shadowing and on-board antenna placement and characteristics can influence the received power strength.

In addition, AA channels between airborne UAVs mostly experience strong LoS similar to the high-altitude AG channels. However, Doppler shift is higher because UAV mobility is significantly higher and it is difficult to maintain alignment between multiple UAVs.

Accurate AG and AA propagation channel models are imperative for the optimal deployment and the design of the UAV communication networks. This section will discuss recent efforts in the modeling of AG and AA propagation channels.

2.4.1 Background

In wireless communications, several propagation phenomena occur when electromagnetic waves radiate from the transmitter in several directions and interact with the surrounding environment before reaching the receiver. As shown in Figure 2.3, propagation phenomena such as reflection, scattering, diffraction, and penetration occur due to the natural obstacles and buildings, which provoke the multiple realization of the signal transmitted from the UAV, often known as multipath components (MPC). Thus, each component received at the receiver with different amplitude, phase, and delay, and the resultant signal is a superposition of multiple copies of the transmitted signal, which can interfere either constructively or destructively depending

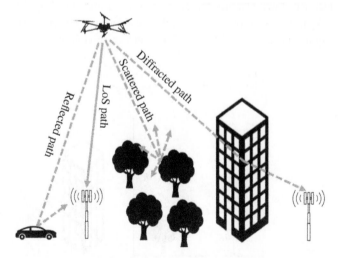

Figure 2.3 Multipath air-to-ground propagation in urban setting.

on their respective random phases [6]. Typically, several fading mechanisms are added linearly in dB to represent the radio channel as

$$y = PL + X_L + X_S, \tag{2.1}$$

where PL is the distance-dependent path loss, X_L is the large-scale fading consisting of power variation on a large scale due to the environment, and X_S is the small-scale fading. Parameters of channel model, such as path loss exponent and LoS probability, are dependent on the altitude level because propagation conditions change at different altitudes. The airspace is often segregated into three propagation echelons or slices as follows:

- *Terrestrial channel*: For suburban and urban environments, altitude is between 10 and 22.5 m, respectively [7]. In this case, the terrestrial channel models can be used to model AG propagation because the airborne UAV is below the rooftop level. As a result, NLoS is the dominant component in the propagation.
- *Obstructed AG channel*: For suburban and urban environments, altitude is 10–40 m and 22.5–100 m, respectively. In this case, LoS probability is higher than that of the terrestrial channels.
- *High-altitude AG channel*: All channels are in LoS for the altitude ranges between 100 and 300 m or above. Consequently, the propagation is similar to that in the free space case. Moreover, no shadowing is experienced for these channels.

2.4.1.1 Path Loss and Large-Scale Fading

Air-to-Air Channel Free space path loss model is the simplest channel model to represent the AA propagation at a relatively high altitude. Thus, the received power is given by [6]

$$P_R = P_T G_T G_R \left(\frac{\lambda_c}{4\pi d} \right)^2, \tag{2.2}$$

where P_T denotes the transmitted signal power, G_T and G_R represent the gain of the transmitter and receiver antennas, respectively, d is the ground distance between the transmitter and receiver, and λ_c is the carrier wavelength. Path loss exponent (η) is the rate of distance-dependent power loss, where η varies with environments. In Eq. (2.2), $\eta = 2$ for free space propagation. Therefore, the distance-dependent path loss expression can be generalized as

$$PL = \left(\frac{4\pi d}{\lambda_c} \right)^\eta. \tag{2.3}$$

Air-to-Ground Channel In urban environment, the AG channel may not experience complete free space propagation. In the existing literature on UAV communications, the log-distance model is the prominently used path loss model due to its simplicity and applicability when environmental parameters are difficult to define. Therefore, path loss in dB is given by

$$PL(d) = PL_0 + 10\eta \log\left(\frac{d}{d_0}\right), \tag{2.4}$$

where $PL_0 = 20 \log\left(\frac{4\pi d_0}{\lambda_c}\right)$ is the path loss for the reference distance d_0. For the same propagation distance between the ground device and the UAV, large-scale variations are different at different locations within the same environment because the materials of obstacles vary from each other, which affects the radio signal propagation. As a result, at any distance d, X_L in Eq. (2.1) is the shadow fading measured in dB and modeled as the normal random variable with variance σ in dB. This model is extensively applied for modeling of the terrestrial channels. Table 2.2 lists some measurement campaigns for the estimations of path loss and large-scale effects.

Another popular channel model to characterize the AG propagation in UAV communications is the probabilistic path loss model in [4] and [17]. In [17], the path loss between the ground device and the UAV is dependent on the position of the UAV and the propagation environments (e.g. suburban, urban, dense-urban, high-rise). Consequently, during the AG radio propagation, the communication link can be either LoS or NLoS depending on the environment. Many of the existing works [18–35] on UAV communications adopted the probabilistic path loss model of [4] and [17]. In these works, the probability of occurrence of LoS and NLoS links are functions of the environmental parameters, height of the buildings, and the elevation angle between the ground device and the UAV. This model is based on environmental parameters defined in the recommendations of the International Telecommunication Union (ITU). In particular, ITU-R provides statistical parameters related to the environment that determine the height, number, and density of the buildings or obstacles. For instance, in [36], the height of the buildings can be modeled by using the Rayleigh distribution. The average path loss for the AG propagation in [17] is given as

$$\overline{PL} = \mathbb{P}_{LoS} \times PL_{LoS} + \left(1 - \mathbb{P}_{LoS}\right) \times PL_{NLoS}, \tag{2.5}$$

where PL_{LoS} and PL_{NLoS} are the LoS and NLoS path loss, respectively, for the free space propagation. $\mathbb{P}_{LoS}$ is the LoS probability given as

$$\mathbb{P}_{LoS} = \frac{1}{1 + \mathcal{A}e^{-B(\theta - \mathcal{A})}}, \tag{2.6}$$

Table 2.2 Measurement campaigns to characterize the path loss and large-scale AG propagation fading.

References	Scenario	η	PL_0 (dB)	σ (dB)
Yanmaz et al. [8]	Urban/Open field	2.2–2.6	—	—
Yanmaz et al. [9]	Open field	2.01	—	—
Ahmed et al. [10]	—	2.32	—	—
Khawaja et al. [11]	Suburban/Open field	2.54–3.037	21.9–34.9	2.79–5.3
Newhall et al. [12]	Urban/Rural	4.1	—	5.24
Tu and Shimamoto [13]	Near airports	2–2.25	—	—
Matolak and Sun [14]	Suburban	1.7 (L-band)	98.2–99.4 (L-band)	2.6–3.1 (L-band)
		1.5–2 (C-band)	110.4–116.7 (C-band)	2.9–3.2 (C-band)
Sun and Matolak [15]	Mountains	1–1.8	96.1–123.9	2.2–3.9
Meng and Lee [16]	Over sea	1.4–2.46	19–129	—

where $\mathcal{A}$ and $\mathcal{B}$ are the constant values related to the environment, $\theta = \arctan\left(\frac{h}{d}\right)$ is the elevation angle between the ground user and the UAV, h is the altitude of the UAV, and d is the distance between the ground projection of the UAV and the ground device. According to Eq. (2.6), as the elevation angle increases with the UAV altitude, the blockage effect decreases and the AG propagation becomes more LoS. An advantage of this model is that it is applicable for different environments and for different UAV altitudes. However, it is unable to capture the impact of path loss for AG propagation in mountainous regions and over water bodies due to the lack of information related to their statistical parameters.

Conventional well-known channel models for cellular communications can be used for UAV communications for UAV altitude between 1.5 and 10 m. One such model for the macro-cell network was designed for the rural environment by the 3rd Generation Partnership Project (3GPP) in [7, 37].

Since LoS and NLoS links are treated separately, the probability of LoS propagation is expressed as

$$\mathbb{P}^G_{\text{LoS}} = \begin{cases} 1, & \text{if } d \leq 10\,\text{m}, \\ e^{-\frac{d-10}{1000}}, & \text{if } 10\,\text{m} < d. \end{cases} \tag{2.7}$$

Path loss and large-scale fading can be calculated once the LoS probability is known from Eq. (2.7). As the communication nodes change their position, path loss also changes and can be found as

$$\text{PL}^G_{\text{LoS}} = \begin{cases} \text{PL}^G_1, & \text{if } 10\,\text{m} \leq d \leq \hat{d}, \\ \text{PL}^G_2, & \text{if } \hat{d} \leq d \leq 10\,\text{km}, \end{cases} \tag{2.8}$$

$$\text{PL}^G_{\text{NLoS}} = \max\left(\text{PL}^G_{\text{LoS}}, \widehat{\text{PL}}^G_{\text{NLoS}}\right), \quad \text{for } 10\,\text{m} \leq d \leq 5\,\text{km}, \tag{2.9}$$

where

$$\text{PL}^G_1 = 20\log\left(\frac{40\pi d f_c}{3}\right) + \min\left(0.03h^{1.72}, 10\right)\log(d) \\ - \min\left(0.44h^{1.72}, 14.77\right) + 0.002d\log(h), \tag{2.10}$$

$$\text{PL}^G_2 = \text{PL}^G_1 + 40\log\left(\frac{d}{\hat{d}}\right), \tag{2.11}$$

$$\widehat{\text{PL}}^G_{\text{NLoS}} = 161.04 - 7.1\log(w) + 7.5\log(h) - \left(24.37 - 3.7\left(\frac{h}{h_G}\right)^2\right)\log(h_G) \\ + \left(43.42 - 3.1\log\left(h_G\right)\right)\left(\log(d) - 3\right) + 20\log(f_c) \\ - \left(3.2\log\left(11.75h\right)^2 - 4.97\right), \tag{2.12}$$

$$\hat{d} = 2\pi h h_G \frac{f_c}{c}, \tag{2.13}$$

with f_c, h_G, w, and c being the carrier frequency, height of ground BS, the average width of street, and the speed of light, respectively.

For the obstructed AG propagation with the UAV altitude between 10 and 40 m, the LoS probability in the rural environment for the macro-cell network can be computed as [7]

$$\mathbb{P}_{\text{LoS}}^{\text{A}} = \begin{cases} 1, & \text{if } d \leq \tilde{d}, \\ \dfrac{\tilde{d}}{d} + e^{\left(\frac{-d}{p_1}\right)\left(1-\frac{\tilde{d}}{d}\right)}, & \text{if } \tilde{d} < d, \end{cases} \tag{2.14}$$

where

$$\tilde{d} = \max\left(1350.8 \log(h) - 1602, 18\right), \tag{2.15}$$

$$p_1 = \max\left(15021 \log(h) - 160\,53, 1000\right). \tag{2.16}$$

The path loss for LoS and NLoS links can be computed as

$$\text{PL}_{\text{LoS}}^{\text{A}} = \max\left(23.9 - 1.8 \log(h), 20\right) \log(d) + 20 \log\left(\frac{40\pi f_c}{3}\right), \tag{2.17}$$

$$\text{PL}_{\text{NLoS}}^{\text{A}} = \max\left(\text{PL}_{\text{LoS}}^{\text{A}}, -12 + (35 - 5.3 \log(h)) \log(d) + 20 \log\left(\frac{40\pi f_c}{3}\right)\right). \tag{2.18}$$

For a high-altitude AG channel with $40\,\text{m} < h \leq 300\,\text{m}$, the LoS probability is 1 and the path loss can be formulated as Eq. (2.17).

2.4.1.2 Small-Scale Fading

Small-scale fading refers to the random fluctuations of amplitude and phase of the received signal over a short distance or a short period of time due to constructive or destructive interference of the MPC. For different propagation environments and wireless systems, different distribution models are suggested to analyze the random variations in the received signal envelop. The Rician and Rayleigh distributions are widely used models in the literature of wireless communications, where both are based on the central limit theorem. The Rician distribution provides better fit for the AA and AG channels, where the impact of LoS propagation is stronger. On the other hand, when the MPC impinges at the receiver with random amplitude and phase, the small-scale fading effect can be captured by the Rayleigh distribution [6].

Geometrical analysis, numerical simulations, and empirical data are used to obtain the stochastic fading models [38–40]. Geometry-based stochastic

Table 2.3 Measured small-scale fading of AG propagation in different environments.

References	Scenario	Frequency band	Fading distribution
Khawaja et al. [11]	Suburban/Open field	Ultra-wideband	Nakagami
Newhall et al. [12]	Urban/Suburban	Wideband	Rayleigh, Rician
Tu and Shimamoto [13]	Urban/Suburban	Wideband	Rician
Matolak and Sun [14]	Urban/Suburban	Wideband	Rician
Simunek et al. [45]	Urban/Suburban	Narrowband	Rician
Cid et al. [46]	Forest/Foliage	Ultra-wideband	Rician, Nakagami
Matolak and Sun [47]	Sea/Fresh water	Wideband	Rician

channel model (GBSCM) is the most popular type of small-scale fading model. GBSCM is subdivided into regular-shaped geometry-based stochastic channel model (RS-GBSCM) and irregular-shaped geometry-based stochastic channel model (IS-GBSCM). Time-variant IS-GBSCM was presented in [41] and RS-GBSCM was presented in [42] and [43].These works illustrated Rician distribution for small-scale fading. In [44], non-geometric stochastic channel model (NGSCM) was provided, where small-scale effects of AG propagation were modeled by using Rician and Loo models. Table 2.3 provides the measured characteristics of small-scale fading of AG propagation in different environments.

2.4.1.3 Airframe Shadowing

Airframe shadowing occurs when the LoS of AG propagation is obstructed by the UAV structure. This impairment is unique to UAV communications for both AA and AG channels and does not exist in conventional cellular communications. Airframe shadowing is more severe in fixed-wing UAVs mounted with single antenna. In this case, the AG communication link can be severe during roll, pitch, or yaw motion of the UAV. One possible solution to alleviate airframe shadowing is to replace the single-antenna system with spatially separated multiple antennas. Other factors responsible for airframe shadowing are the size, shape, and material of the UAV. The seminal work on the measurement of airframe shadowing was performed in [48], which found that the aircraft roll angle was proportional to the shadowing attenuation. Moreover, shadowing duration depends on the flight maneuvering.

2.5 Key Research Challenges of UAV-Enabled Wireless Network

This section discusses some of the key research challenges for the practical deployment of UAVs as airborne wireless nodes.

2.5.1 Optimal Deployment of UAVs

In UAV-based communications, one of the key challenges is the optimal three-dimensional deployment of hovering UAV. The capability of UAV to maneuver and adjust its altitude provides additional degree of freedom for UAV deployment in an efficient manner to improve capacity and coverage. In fact, UAV deployment is more challenging in UAV communications than in conventional terrestrial communications because the characteristics of AG propagation change with the position of the UAV. However, for efficient UAV deployment, flight duration and energy constraints must be taken into account for battery-operated UAV, as they affect the performance of networks. In addition, simultaneous deployment of multiple UAVs is more challenging because of the co-channel interference and the possibility of airborne collision of UAVs. Another important issue is the UAV deployment in the presence of terrestrial network. UAV deployment problem has been extensively discussed in the literature for coverage maximization [17, 29, 30, 33, 33], data collection from Internet of Things (IoT) devices [31], UAV-assisted wireless network [27], disaster scenario [49], and caching applications [22].

2.5.2 UAV Trajectory Optimization

Optimal trajectory design for mobile UAV is an important issue in UAV-based communications. Specifically, optimal path planning is crucial for UAVs operating for data collection from ground-based sensors and caching scenarios. UAV trajectory planning is mostly effected by the dimension of the target area, flight duration of the mission, QoS requirement by the ground users, and energy constraints. Apart from physical parameters, UAV trajectory optimization is analytically a challenging problem because it involves a fixed number of optimization variables related to the UAV locations [1]. In addition, UAV trajectory optimization requires coupling between different QoS metrics in wireless communication with the mobility of UAV. Recently, there have been a number of studies on the joint trajectory optimization of UAV with its wireless communication metrics, such as throughput maximization in [50–52] and energy-efficient UAV communication in [53, 54].

2.5.3 Energy Efficiency and Resource Management

Energy efficiency and resource management require attention where UAVs are operating in key scenarios to collect data from IoT devices, ensure public safety, and support cellular wireless network. Resource management is a major challenge in UAV communications unlike in cellular communications [55]. However, UAV communications introduce additional hindrance in radio resource management due to the interplay between the UAV flight duration, mobility pattern, limited energy source, and spectral efficiency. Therefore, in [56], resource management was jointly optimized with the UAV trajectory in wireless environment.

Limited amount of on-board energy is available for battery-operated UAV, which must be used for propulsion and to fulfill communication-related tasks [5]. Consequently, continuous and long-term wireless coverage curtails the UAV flight time. In addition, UAV energy consumption also depends on its path, weather condition, and mission of the UAV. Thus, energy constraints of UAV must be explicitly taken into account during planning of the UAV-based communication systems. Various works have studied the interplay between energy efficiency and the optimal UAV trajectory [53–55].

2.6 Conclusion

This chapter discussed the use of UAVs in wireless communication network, specifically, the use of UAVs as aerial BSs and as aerial UE in cellular-assisted systems. In both cases, the accurate channel model of the AG and AA propagation is paramount, which must take into account the environmental conditions, wireless channel impairments, and the UAV mobility to characterize the performance of UAV-based communication network. Some channel modeling efforts have been studied in this chapter. In addition, key challenges, such as optimal deployment of UAVs, optimization of trajectory path, resource management, and energy efficiency, have also been highlighted.

Bibliography

1 Zeng, Y., Zhang, R., and Lim, T.J. (2016). Wireless communications with unmanned aerial vehicles: opportunities and challenges. *IEEE Communications Magazine* 54 (5): 36–42.

2 Qualcomm Technologies Inc. (2016). Leading the World to 5G: Evolving Cellular Technologies for Safer Drone Operation. Technical report. Qualcomm.

3 Patterson, T. (2015). Google, Facebook, SpaceX, OneWeb plan to beam internet everywhere. https://edition.cnn.com/2015/10/30/tech/pioneers-google-facebook-spacex-oneweb-satellite-drone-balloon-internet/index.html (accessed 08 March 2021).

4 Al-Hourani, A., Kandeepan, S., and Jamalipour, A. (2014). Modeling air-to-ground path loss for low altitude platforms in urban environments. *2014 IEEE Global Communications Conference*, pp. 2898–2904.

5 Fotouhi, A., Qiang, H., Ding, M. et al. (2019). Survey on UAV cellular communications: practical aspects, standardization advancements, regulation, and security challenges. *IEEE Communications Surveys Tutorials* 21 (4): 3417–3442.

6 Molisch, A. (2011). *Wireless Communications*. Wiley - IEEE.

7 3GPP (2017). Study on Enhanced LTE Support for Aerial Vehicles. Technical report, *3rd Generation Partnership Project 3GPP*.

8 Yanmaz, E., Kuschnig, R., and Bettstetter, C. (2011). Channel measurements over 802.11a-based UAV-to-ground links. *2011 IEEE GLOBECOM Workshops (GC Wkshps)*, pp. 1280–1284.

9 Yanmaz, E., Kuschnig, R., and Bettstetter, C. (2013). Achieving air-ground communications in 802.11 networks with three-dimensional aerial mobility. *2013 Proceedings IEEE INFOCOM*, pp. 120–124.

10 Ahmed, N., Kanhere, S.S., and Jha, S. (2016). On the importance of link characterization for aerial wireless sensor networks. *IEEE Communications Magazine* 54 (5): 52–57.

11 Khawaja, W., Guvenc, I., and Matolak, D. (2016). UWB channel sounding and modeling for UAV air-to-ground propagation channels. *2016 IEEE Global Communications Conference (GLOBECOM)*, pp. 1–7.

12 Newhall, W.G., Mostafa, R., Dietrich, C. et al. (2003). Wideband air-to-ground radio channel measurements using an antenna array at 2 GHz for low-altitude operations. *IEEE Military Communications Conference, 2003. MILCOM 2003*, Volume 2, pp. 1422–1427.

13 Tu, H.D. and Shimamoto, S. (2009). A proposal of wide-band air-to-ground communication at airports employing 5-GHz band. *2009 IEEE Wireless Communications and Networking Conference*, pp. 1–6.

14 Matolak, D.W. and Sun, R. (2017). Air-ground channel characterization for unmanned aircraft systems-part III: the suburban and near-urban environments. *IEEE Transactions on Vehicular Technology* 66 (8): 6607–6618.

15 Sun, R. and Matolak, D.W. (2017). Air-ground channel characterization for unmanned aircraft systems part II: Hilly and mountainous settings. *IEEE Transactions on Vehicular Technology* 66 (3): 1913–1925.

16 Meng, Y.S. and Lee, Y.H. (2011). Measurements and characterizations of air-to-ground channel over sea surface at C-band with low airborne altitudes. *IEEE Transactions on Vehicular Technology* 60 (4): 1943–1948.

17 Al-Hourani, A., Kandeepan, S., and Lardner, S. (2014). Optimal lap altitude for maximum coverage. *IEEE Wireless Communications Letters* 3 (6): 569–572.

18 Bor-Yaliniz, R.I., El-Keyi, A., and Yanikomeroglu, H. (2016). Efficient 3-D placement of an aerial base station in next generation cellular networks. *2016 IEEE International Conference on Communications (ICC)*, pp. 1–5.

19 Bor-Yaliniz, I. and Yanikomeroglu, H. (2016). The new frontier in ran heterogeneity: multi-tier drone-cells. *IEEE Communications Magazine* 54 (11): 48–55.

20 Hayajneh, A.M., Zaidi, S.A.R., McLernon, D.C., and Ghogho, M. (2016). Drone empowered small cellular disaster recovery networks for resilient smart cities. *2016 IEEE International Conference on Sensing, Communication and Networking (SECON Workshops)*, pp. 1–6.

21 Gomez, K., Hourani, A., Goratti, L. et al. (2015). Capacity evaluation of aerial LTE base-stations for public safety communications. *2015 European Conference on Networks and Communications (EuCNC)*, pp. 133–138.

22 Chen, M., Mozaffari, M., Saad, W. et al. (2017). Caching in the sky: proactive deployment of cache-enabled unmanned aerial vehicles for optimized quality-of-experience. *IEEE Journal on Selected Areas in Communications* 35 (5): 1046–1061.

23 Challita, U. and Saad, W. (2017). Network formation in the sky: unmanned aerial vehicles for multi-hop wireless backhauling. *GLOBECOM 2017 - 2017 IEEE Global Communications Conference*, pp. 1–6.

24 Kalantari, E., Yanikomeroglu, H., and Yongacoglu, A. (2016). On the number and 3D placement of drone base stations in wireless cellular networks. *2016 IEEE 84th Vehicular Technology Conference (VTC-Fall)*.

25 Shakhatreh, H., Khreishah, A., Chakareski, J. et al. (2016). On the continuous coverage problem for a swarm of UAVs. *2016 IEEE 37th Sarnoff Symposium*, pp. 130–135.

26 Azari, M.M., Rosas, F., Chen, K., and Pollin, S. (2016). Joint sum-rate and power gain analysis of an aerial base station. *2016 IEEE Globecom Workshops (GC Wkshps)*, pp. 1–6.

27 Hayajneh, A.M., Zaidi, S.A.R., McLernon, D.C., and Ghogho, M. (2016). Optimal dimensioning and performance analysis of drone-based wireless communications. *2016 IEEE Globecom Workshops (GC Wkshps)*, pp. 1–6.

28 Jia, S. and Zhang, L. (2017). Modelling unmanned aerial vehicles base station in ground-to-air cooperative networks. *IET Communications* 11 (8): 1187–1194.

29 Mozaffari, M., Saad, W., Bennis, M., and Debbah, M. (2016). Efficient deployment of multiple unmanned aerial vehicles for optimal wireless coverage. *IEEE Communications Letters* 20 (8): 1647–1650.

30 Mozaffari, M., Saad, W., Bennis, M., and Debbah, M. (2015). Drone small cells in the clouds: design, deployment and performance analysis. In *2015 IEEE Global Communications Conference (GLOBECOM)*, pp. 1–6.

31 Mozaffari, M., Saad, W., Bennis, M., and Debbah, M. (2017). Mobile unmanned aerial vehicles (UAVs) for energy-efficient internet of things communications. *IEEE Transactions on Wireless Communications* 16 (11): 7574–7589.

32 Azari, M.M., Rosas, F., Chen, K., and Pollin, S. (2018). Ultra reliable UAV communication using altitude and cooperation diversity. *IEEE Transactions on Communications* 66 (1): 330–344.

33 Alzenad, M., El-Keyi, A., Lagum, F., and Yanikomeroglu, H. (2017). 3-D placement of an unmanned aerial vehicle base station (UAV-BS) for energy-efficient maximal coverage. *IEEE Wireless Communications Letters* 6 (4): 434–437.

34 Alzenad, M., El-Keyi, A., and Yanikomeroglu, H. (2018). 3-D placement of an unmanned aerial vehicle base station for maximum coverage of users with different QoS requirements. *IEEE Wireless Communications Letters* 7 (1): 38–41.

35 Khuwaja, A.A., Zheng, G., Chen, Y., and Feng, W. (2019). Optimum deployment of multiple UAVs for coverage area maximization in the presence of co-channel interference. *IEEE Access* 7: 85203–85212.

36 International Telecommunication Union (ITU) (2003). Propagation Data and Prediction Methods for the Design of Terrestrial Broadband Millimetric Radio Access Systems. Technical report. International Telecommunication Union (ITU).

37 3GPP (2018). Study on Channel Model for Frequencies from 0.5 to 100 GHz. Technical report. 3rd Generation Partnership Project (3GPP).

38 Wentz, M. and Stojanovic, M. (2015). A MIMO radio channel model for low-altitude air-to-ground communication systems. *2015 IEEE 82nd Vehicular Technology Conference (VTC2015-Fall)*, pp. 1–6.

39 Gulfam, S.M., Nawaz, S.J., Ahmed, A., and Patwary, M.N. (2016). Analysis on multipath shape factors of air-to-ground radio communication channels. *2016 Wireless Telecommunications Symposium (WTS)*, pp. 1–5.

40 Zeng, L., Cheng, X., Wang, C., and Yin, X. (2017). Second order statistics of non-isotropic UAV ricean fading channels. *2017 IEEE 86th Vehicular Technology Conference (VTC-Fall)*, pp. 1–5.

41 Blandino, S., Kaltenberger, F., and Feilen, M. (2015). Wireless channel simulator testbed for airborne receivers. *2015 IEEE Globecom Workshops (GC Wkshps)*, pp. 1–6.

42 Ksendzov, A. (2016). A geometrical 3D multi-cluster mobile-to-mobile MIMO channel model with rician correlated fading. *2016 8th International Congress on Ultra Modern Telecommunications and Control Systems and Workshops (ICUMT)*, pp. 191–195.

43 Gao, X., Chen, Z., and Hu, Y. (2013). Analysis of unmanned aerial vehicle MIMO channel capacity based on aircraft attitude. *WSEAS Transactions on Information Science and Applications* 10 (2): 58–67.

44 Yang, J., Liu, P., and Mao, H. (2011). Model and simulation of narrowband ground-to-air fading channel based on Markov process. *2011 International Conference on Network Computing and Information Security*, Volume 1, pp. 142–146. https://doi.org/10.1109/NCIS.2011.37.

45 Simunek, M., Fontán, F.P., and Pechac, P. (2013). The UAV low elevation propagation channel in urban areas: statistical analysis and time-series generator. *IEEE Transactions on Antennas and Propagation* 61 (7): 3850–3858.

46 Cid, E.L., Alejos, A.V., and Sanchez, M.G. (2016). Signaling through scattered vegetation: empirical loss modeling for low elevation angle satellite paths obstructed by isolated thin trees. *IEEE Vehicular Technology Magazine* 11 (3): 22–28.

47 Matolak, D.W. and Sun, R. (2016). Air-ground channels for UAS: summary of measurements and models for L- and C-bands. *2016 Integrated Communications Navigation and Surveillance (ICNS)*, p. 8B2-1–8B2-11.

48 Sun, R., Matolak, D.W., and Rayess, W. (2017). Air-ground channel characterization for unmanned aircraft systems-part IV: airframe shadowing. *IEEE Transactions on Vehicular Technology* 66 (9): 7643–7652.

49 Kosmerl, J. and Vilhar, A. (2014). Base stations placement optimization in wireless networks for emergency communications. *2014 IEEE International Conference on Communications Workshops (ICC)*, pp. 200–205.

50 Valiulahi, I. and Masouros, C. (2020). Multi-UAV deployment for throughput maximization in the presence of co-channel interference. *IEEE Internet of Things Journal* 8 (5: 3605–3618.

51 Zhao, N., Pang, X., Li, Z. et al. (2019). Joint trajectory and precoding optimization for UAV-assisted NOMA networks. *IEEE Transactions on Communications* 67 (5): 3723–3735.

52 Zeng, Y., Zhang, R., and Lim, T.J. (2016). Throughput maximization for UAV-enabled mobile relaying systems. *IEEE Transactions on Communications* 64 (12): 4983–4996.

53 Zeng, Y. and Zhang, R. (2017). Energy-efficient UAV communication with trajectory optimization. *IEEE Transactions on Wireless Communications* 16 (6): 3747–3760.

54 Yang, Z., Pan, C., Wang, K., and Shikh-Bahaei, M. (2019). Energy efficient resource allocation in UAV-enabled mobile edge computing networks. *IEEE Transactions on Wireless Communications* 18 (9): 4576–4589.

55 Mumtaz, S., Huq, K.M.S., Radwan, A. et al. (2014). Energy efficient interference-aware resource allocation in LTE-D2D communication. *2014 IEEE International Conference on Communications (ICC)*, pp. 282–287.

56 Cui, F., Cai, Y., Qin, Z. et al. (2019). Multiple access for mobile-UAV enabled networks: joint trajectory design and resource allocation. *IEEE Transactions on Communications* 67 (7): 4980–4994.

3

Ultra-wideband Channel Measurements and Modeling for Unmanned Aerial Vehicle-to-Wearables (UAV2W) Systems

Amit Kachroo[1], Surbhi Vishwakarma[1], Jacob N. Dixon[2], Hisham Abuella[1], Adithya Popuri[1], Qammer H. Abbasi[3], Charles F. Bunting[1], Jamey D. Jacob[4], and Sabit Ekin[1]

[1] School of Electrical and Computer Engineering, Oklahoma State University, Stillwater, OK, USA
[2] IBM, Rochester, MN, USA
[3] School of Engineering, University of Glasgow, Glasgow, UK
[4] School of Mechanical and Aerospace Engineering, Oklahoma State University, Stillwater, OK, USA

3.1 Introduction

Over the past decades, wireless technology has seen an upward trend in the bandwidth of the signals employed. The main reason behind this upward trend is the proliferation of multimedia technologies that demand high data rate, and also an increase in user base. Ultra-wideband (UWB) radio is one of the creations of this trend where the bandwidth occupied by UWB technology is greater than or equal to 500 MHz. Therefore, UWB communication technology exploits this large bandwidth to catch up with the high data rate. Apart from the high bandwidth, the main advantages of UWB can be listed as follows:

- Low power consumption with high data rate. The received power in UWB lies very close to the noise floor [1–5].
- Control over duty cycle makes the battery last longer.
- Low probability of detection as it is close to the noise floor and any attempt of jamming or eavesdropping will make the signal noisy [6].
- Small wavelength with low power makes it a perfect fit for body-centric wireless network [3, 4].

Given these advantages, UWB is best suited for off-body communication. Moreover, the Federal Communication Commission's (FCC) guideline of the

Autonomous Airborne Wireless Networks, First Edition.
Edited by Muhammad Ali Imran, Oluwakayode Onireti, Shuja Ansari, and Qammer H. Abbasi.
© 2021 John Wiley & Sons Ltd. Published 2021 by John Wiley & Sons Ltd.

power limit of −41.3 dBm or 75 nW/MHz identifies the UWB technology as an unintentional interference source; the fact that it can thereby coexist with other wireless technologies, especially at 2.4 GHz (WiFi, Bluetooth) with minimal or no interference, reinforces the application of UWB technologies for off-body communication.

On the other hand, unmanned aerial vehicles (UAVs) are being now used for remote healthcare deliveries especially to far flung areas that lack connectivity. UAVs are also being used for emergency medical deliveries where time is of utmost importance, such as during cardiac arrests [7–10]. One of the upcoming themes for UAVs is to directly monitor the health of a patient by utilizing wearable patch devices [1, 7, 10–12]. The study in this chapter explores the UWB technology with UAVs further for health monitoring applications. This type of setup involving UAV and wearable antenna/antennas is also known as unmanned aerial vehicle-to-wearable (UAV2W) systems [1].

The closest one to this study is our previous work [1], where different UWB bandwidths were considered for channel modeling in an indoor environment. However, in this work, we consider the complete UWB bandwidth of 7.5 GHz to study these body channels, and also to look into two different environments and study the effect of postures. Also, previous studies such as [5, 13–15] have performed on-body radio channel characterization and modeling at 2.45 GHz but not at the UWB frequency. In addition [16–18] performed off-body radio channel studies in a contained scenario with antennas placed in standalone position. The other closest study is in [2, 3], where off- and on-body channel characterizations are performed without the real human subject. To the best of our knowledge, this is one of the first works to consider UWB channel characterization between humans and UAV at 7.5 GHz bandwidth, and has studied different environments and the effects of different body postures on the UWB system.

The rest of the chapter is organized as follows: Section 3.2 discusses the measurement setup and data acquisition part, and Section 3.3 covers the UWB-UAV2W radio channel characterization. Section 3.4 details the statistical analysis and finally, Section 3.5 presents the conclusion based on the measurement campaign done so far.

3.2 Measurement Settings

There are generally two methods to measure the channel response in a wireless communication, either time correlator based or frequency sweep based. In our work, we have utilized the latter one by using a Vector Network

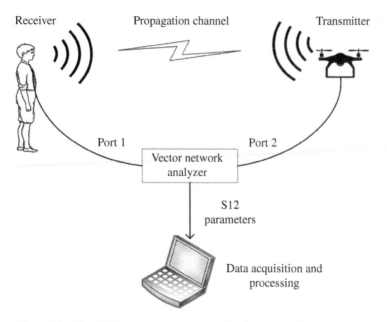

Figure 3.1 The UWB measurement communication setup. Source: Kachroo et al. [1] With permission of IEEE.

Analyzer (VNA). Figure 3.1 shows the measurement setup considered in this chapter. In this setup, the receiver (Rx) is patched at specific location on a human subject with the transmitter (Tx) antenna on a drone that hovers at a fixed height, and at various distances from the human subject. The Rx and Tx antennae are connected at port 1 and port 2 of the VNA respectively. The transmit power of the VNA is kept at −40 dB so that it is under the FCC regulations involving humans as mentioned in the Section 3.1.[1] The relationship of noise floor with bandwidth is given as

$$\text{Noise floor (dBm)} = 10\log_{10}(kT) + 10\log_{10}(B) + \text{NF}, \tag{3.1}$$

where k is the Boltzmann's constant, T is the temperature (K) at the receiver, B is the bandwidth under consideration, and NF is the noise figure at the receiver. In our measurement settings, with a bandwidth of 7.5 GHz, a transmit power of −40 dB is considered to be tolerable.

The VNA was given a frequency sweep from 3.1 to 10.6 GHz with 1601 uniformly distributed points in the frequency domain. The VNA was then calibrated with the 2-port calibration mechanism (open-short-load-through)

---

1 No power amplifier was used at the transmitter because of the human subject regulations from FCC.

to remove the cable effects from the measurements. The measurement data, that is, the S12 parameters, will represent the channel transfer function in the frequency domain. The S12 parameters thus acquired are processed by Python and MATLAB scripts on a laptop. By taking the inverse fast Fourier transform (IFFT), various time domain analyses such as root mean square (RMS) time and maximum excess time will be also analyzed.

The specifications of all the equipment used in this measurement campaign are shown in Table 3.1.

The UWB antenna (Octane BW-3000-10000-EG) is an omnidirectional wideband antenna with 5.5 dBi, 8.2 dBi, and 6.3 dBi gains at 3 GHz, 6 GHz, and 9 GHz. This lightweight (2 Oz) antenna is also small in size with dimensions of $4.5'' \times 4.25'' \times 0.4''$ and voltage standing wave ratio of less than $2 : 1$. The IRIS+ quadcopter utilized in this measurement campaign has a three DR link for communication and a maximum speed of 25 mph.[2] The patch antenna and the UAV can be visualized from Figure 3.2.

Table 3.1 The measurement apparatus with their specifications.

Equipment	Specifications
Vector network analyzer	Agilent 8722ES (50 MHz–40 GHz)
Calibration kit	Agilent 85032F
UAV	3 DR IRIS + Quadcopter
Two antenna sensors	Octane BW-3000 (3–11 GHz)

(a)   (b)

Figure 3.2 The UWB antenna and the IRIS+ quadcopter used in the measurement campaign. (a) Octane antenna. (b) IRIS+ quadcopter. Source: Kachroo et al. [1] With permission of IEEE.

---

2 For more information, please check this link – [19].

The UWB antenna for the measurement was placed at nine different body locations for the line-of-sight (LoS) case, four body positions for the non-line-of-sight (NLoS) case, and two body locations (forehead and abdomen) for different postures. These different body locations in different environments can be visualized from Figure 3.3. The reason for using only four body locations in the NLoS case is that during the LoS measurement campaign, we observed that the measurement for the left arm/left shin is akin to the results of the right arm/right shin due to body symmetry. Hence, for NLoS measurements, four sensor locations (right side) were considered to save time and effort.

As mentioned earlier, the transmitter antenna is placed on the UAV that was airborne at an established height, and the receiver antenna was patched to the human subject; thus, the distance between Tx and Rx was changed by moving the human subject (receiver) 0.5 m to cover 10 points on the ground, which was equivalent to the diagonal distance of 8.0 to 3.5 m toward the UAV. The sketch plan is shown in Figure 3.4 with the marked point and the diagonal distances when the human subject (receiver) is farthest and nearest to the UAV.

At each distance point, the UWB antenna was placed at different body locations and a total of 10 snapshots of the S12 data were recorded, which was later averaged to have a reduced probability of error. The S12 data consists of frequency response in both magnitude (dB) and phase (degrees).

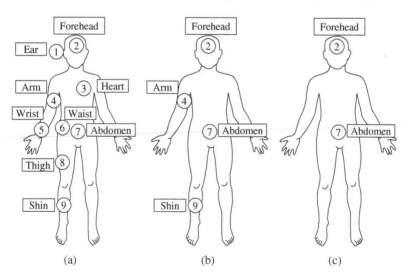

Figure 3.3 The UWB antenna patch locations on the human body for the UWB measurement campaign. (a) Antenna location for LoS scenario. (b) Antenna location for NLoS scenario. (c) Antenna location for different body postures.

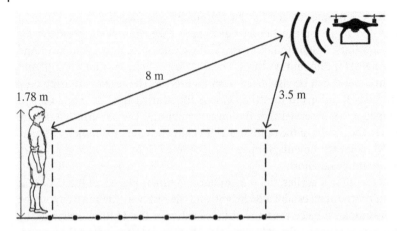

Figure 3.4 The sketch plan of the measurement campaign with the 10 distinct points. The diagonal range varied from 8 meters to 3.5 meters with this layout. Source: Kachroo et al. [1] With permission of IEEE.

For each frequency response, the transfer function was determined after averaging and then taking the IFFT of the transfer function to get the time domain characteristics of the channel. The LoS and NLoS measurements were carried out in the indoor (warehouse) and outdoor environments.

Figure 3.5 shows the actual indoor and outdoor measurement environments for the campaign. Moreover, in the third case of different postures (standing, sleeping, sitting, and bending), the human subject was kept at a fixed distance of 6 m with two crucial static body positions (forehead and abdomen).

In Section 3.3, we will do an in-depth analysis of the S12 data collected in the measurement campaign and characterize UWB channel fading and time dispersion properties for such UAV2W systems.

(a)   (b)

Figure 3.5 Different environments considered for the measurement campaign. (a) Outdoor parking lot environment. (b) Indoor warehouse environment.

3.3 UWB-UAV2W Radio Channel Characterization

In this section, we will look into the next part of the campaign, that is, data processing and analysis. We will analyze the path loss, UWB channel fading characteristics, and time dispersion properties under different environments for different body locations. First, we will do the path loss analysis.

3.3.1 Path Loss Analysis

The frequency response in such a setup is a function of frequency, time, and distance. Therefore, the average path loss at each distance point is obtained by averaging the frequency responses $H(t_i, f_j, d)$ over time and frequency:

$$\text{PL}(d) = \frac{1}{N}\frac{1}{M}\sum_{i=1}^{N}\sum_{i=1}^{M}|H(t_i, f_j, d)|^2 \propto d^\gamma , \qquad (3.2)$$

where $N = 10$, $M = 1601$, and $H(t_i, f_j, d)$ is the channel transfer function (or the S12 parameters). From the literature, the well-known path loss equation with shadowing is given as [20, 21],

$$\text{PL}(d) = \text{PL}(d_0) + 10\gamma\log_{10}\left(\frac{d}{d_0}\right) + X_\sigma(d), \qquad (3.3)$$

where d is the distance between the UWB Tx and Rx that is patched at a specific location on the human body, $\text{PL}(d)$ is the path loss in dB at that distance, d_0 is the reference distance, and $\text{PL}(d_0)$ is the path loss in dB at reference distance d_0. Also, γ is the path loss exponent and $X_\sigma(d)$ is the shadowing factor. The shadowing factor is represented by a Gaussian distribution with zero mean and a variance of σ^2. This equation can be easily solved with Linear regression to determine the path loss exponent. This is utilized for both the data measured in the indoor and outdoor environment under LoS and NLoS scenarios. Figure 3.6 shows one such case of application of linear fit to find the path loss exponent in the indoor environment for the case of UWB radio channel between waist, abdomen, and UAV.

The path loss and path loss exponent values of UWB radio channels in both indoor and outdoor environments for nine body locations in the LoS case are shown in Table 3.2. For analytical simplicity, two or three body locations were combined as a single location and their individual path loss measurements and path loss factor were averaged. The body locations that were merged are wrist/arm as hand, abdomen/waist/heart as chest, ear/forehead as head, and thigh/shin as a leg. Table 3.3 shows the result of the combining of body location with both path loss measurement and path loss exponent.

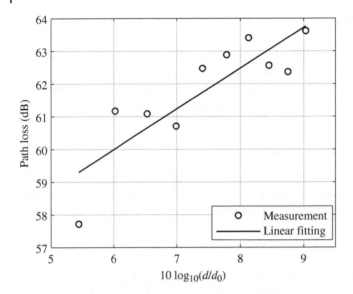

Figure 3.6 Path loss factor determination from linear regression for a wireless channel between waist+forehand and a UAV in an indoor environment.

Similarly, in the case of NLoS, the path loss measurement and path loss factor for the UWB radio channel between different body locations and UAV are given in Table 3.4. It can be clearly inferred from these measurement results that the forehead is the best position for the UWB antenna as it has the lowest path loss while the ear is the worst position with the highest path loss in both LoS and NLoS cases in the indoor and outdoor environments. The main reason behind the ear being the worst case is that the antenna patching on the ear will be in a different direction. This leads to the loss of directivity between the UWB antenna on the ear and an antenna on a UAV (polarization loss); forehead, on the other hand, is the topmost body position with the antenna directed toward the UAV, giving it better directivity. By combining the body parts, the head portion as anticipated is the best position for the UWB antenna placement, while the leg portion results came out to be the worst. Secondly, after path loss, we will shed light on time dispersion analysis, especially the RMS delay, maximum excess delay, and mean delay.

3.3.2 Time Dispersion Analysis

Given the effect of multipath and large-scale fading environment, the UWB channel between different body locations and a UAV will have different delays. In this section, we will analyze three delay parameters, maximum

Table 3.2 Path loss measurement and path loss exponent for nine different body locations in indoor and outdoor environments for the LoS case.

Body location	Indoors	Outdoors
	Path loss measurements (dB)	
Ear	63.88	66.00
Forehead	59.44	60.10
Chest	62.14	60.31
Right arm	62.64	63.17
Right wrist	64.67	63.48
Waist	64.10	64.27
Right thigh	63.89	63.81
Right lower shin	63.52	63.77
Abdomen	62.57	61.96
Overall body average	62.98	62.98
	Path loss exponent	
Ear	0.31	0.57
Forehead	1.80	1.22
Chest	0.19	1.05
Right arm	0.05	0.94
Right wrist	0.34	0.32
Waist	0.96	0.85
Right thigh	0.21	0.58
Right lower shin	0.26	0.13
Abdomen	0.51	0.58
Overall body average	0.51	0.69

excess, mean excess, and RMS delay based on the power delay profile (PDP) [20, 21]. Mathematically, PDP is given as

$$P(t, \tau) = |h(t, \tau)|^2, \tag{3.4}$$

where, $h(t, \tau)$ is the channel response at distance d with delay τ and $P(t, \tau)$ is the measured power. The mean excess delay is the first moment of PDP and is given as [1],

$$\bar{\tau} = \frac{\sum_i^n \tau_i P(\tau_i)}{\sum_i^n P(\tau_i)}. \tag{3.5}$$

Table 3.3 Combined path loss measurement and path loss exponent for four different body locations in indoor and outdoor environments for the LoS case.

Body location	Indoors	Outdoors
Path loss measurements (dB)		
Head	61.66	63.05
Chest	62.9	62.2
Hand	63.65	63.33
Leg	63.70	63.79
Path loss exponent		
Head	1.05	0.89
Chest	0.55	0.82
Hand	0.19	0.63
Leg	0.24	0.36

Table 3.4 Path loss measurement and path loss exponent for four different body locations in indoor and outdoor environments for the NLoS case.

Body location	Indoors	Outdoors
Path loss measurements (dB)		
Forehead	66.98	66.44
Right arm	69.20	68.63
Right lower shin	68.97	69.32
Abdomen	69.32	67.60
Path loss exponent		
Forehead	1.12	0.61
Right arm	0.26	0.38
Right lower shin	0.03	0.15
Abdomen	0.11	0.50

The RMS delay spread is basically the second moment of PDP and is given as

$$\sigma_t = \sqrt{(\overline{\tau^2} - \overline{\tau}^2)}, \tag{3.6}$$

where τ^2 is computed as

$$\tau^2 = \frac{\sum_i^n \tau^2 P(\tau_i)}{\sum_i^n P(\tau_i)}. \tag{3.7}$$

Here n represents the total number of samples. Moreover, the maximum excess delay is the time delay during which the received power falls below a specific defined threshold value. In our case, the threshold was set at 5 dB.

As mentioned earlier, the PDP of a wireless channel describes the average power of the received signal in terms of the delay with respect to the first arrival path in multipath transmission. The PDP with delay spread at different distances can be seen in Figure 3.7, while the comparison of maximum excess, mean excess, and RMS delay in the case of LoS scenario is shown in Figure 3.8. One important point to note here is that as the distance increases between the transmitter and the receiver, so will be the delay of the received signal and its multipath component. This can be easily visualized from Figure 3.7.

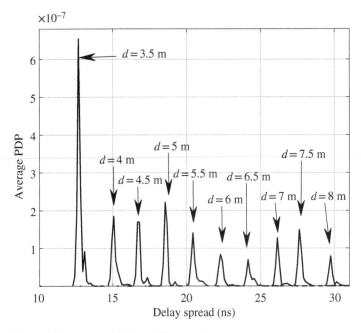

Figure 3.7 Averaged PDP at different distances.

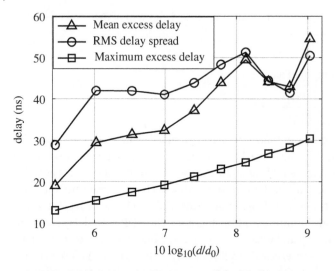

Figure 3.8 Normalized averaged path loss delay comparison.

Similar intuition can be applied with the different delays (mean excess delay, maximum excess delay, and RMS delay) associated with the PDP, that is, longer the distance between the transmitter and receiver, higher would be the corresponding delay value, which can be easily inferred from Figure 3.8.

In addition, Tables 3.5 and 3.6 show the different recorded data measurements of these parameters in different environments with LoS and NLoS cases.

3.3.3 Path Loss Analysis for Different Postures

Finally, we will look into the case of different body postures for path loss and time dispersion characteristics. In the case of different body postures (standing, sitting, bending, and sleeping), the human subject was kept at a fixed distance of 6 m. The environment under consideration was indoor and outdoor with two body locations of forehead and abdomen. Table 3.7 lists the measurement values for these four different body postures in the indoor and outdoor environments.

3.3.4 Time Dispersion Analysis for Different Postures

In this section, we will look into the time dispersion characteristics (mean excess, maximum excess, and RMS delay profile) for the two body locations, both indoors and outdoors, considering four postures. Table 3.8 details all the measurement values for these cases in the indoor and outdoor

Table 3.5 Time dispersion analysis in the case of LoS for nine body locations.

Body patch location	Indoors	Outdoors
RMS delay (ns)		
Ear	43.3	64.2
Forehead	30.5	64.0
Chest	42.9	41.6
Right arm	46.0	60.7
Right wrist	56.8	61.2
Waist	48.6	59.2
Right thigh	49.6	55.1
Right lower shin	48.2	51.5
Abdomen	40.9	58.3
Overall body average	45.2	57.3
Mean excess delay (ns)		
Ear	22	13.6
Forehead	21.7	22.6
Chest	22.5	22.6
Right arm	23.1	12.7
Right wrist	73.5	16
Waist	41	12.7
Right thigh	24.5	25
Right lower shin	40.2	24.9
Abdomen	22.6	68.4
Overall body average	32.34	24.2

environments. From these measurements, considering the RMS delay, forehead is the best position with a delay of 30.5 ns in the indoor environment for the LoS case, and chest is the best location with a delay of 41.6 ns in the outdoor environment for the LoS case while abdomen is the best position with delay of 57.3 and 61.4 ns in both indoor and outdoor environments for the NLoS case, respectively.

Now, considering the mean excess delay, forehead is the best position with delay of 30 ns indoor for the LoS case and also with values of 81.9 and 92.3 ns, respectively, in the NLoS case. However, chest is the best location with delay

Table 3.6 Time dispersion analysis in the case of NLoS for four body locations.

Body patch location	Indoors	Outdoors
RMS delay (ns)		
Forehead	61.0	64.0
Right arm	58.9	62.0
Right lower shin	58.9	62.5
Abdomen	57.3	61.4
Overall body average	59.0	62.4
Mean excess delay (ns)		
Forehead	81.9	92.3
Right arm	107	106
Right lower shin	110	103
Abdomen	110.8	106
Overall body average	102.4	101.8
Maximum excess delay (ns)		
Forehead	135	152
Right arm	213	213
Right lower shin	213	213
Abdomen	209	213
Overall body average	192.5	197.7

Table 3.7 Path loss values in the indoor and outdoor environments for four postures with antenna patched at two body positions.

location	Standing	Bending	Sitting	Sleeping
Indoor environment (dB)				
Forehead	59.20	67.02	60.60	67.84
Abdomen	62.34	65.13	63.34	66.83
Outdoor environment (dB)				
Forehead	59.63	68.30	62.38	68.97
Abdomen	61.48	68.46	60.31	67.53

Table 3.8 Delay analysis values in nanoseconds for two body locations considering four different body postures in the indoor and outdoor environments.

Location	Standing	Bending	Sitting	Sleeping
RMS (indoors)				
Forehead	30.02	61.45	34.46	61.19
Abdomen	38.78	56.36	39.00	61.76
RMS (outdoors)				
Forehead	50.14	63.03	56.40	64.94
Abdomen	57.48	63.86	35.81	61.95
Maximum excess delay (indoors)				
Forehead	21.75	24.84	22.5	23.35
Abdomen	22.28	22.68	23.08	23.08
Maximum excess delay (outdoors)				
Forehead	22.28	24.68	22.81	24.68
Abdomen	22.94	26.01	23.88	23.88
Mean excess delay (indoors)				
Forehead	29.92	70.47	33.53	73.30
Abdomen	36.34	58.57	37.10	75.16
Mean excess delay (outdoors)				
Forehead	45.81	103.00	55.9	92.01
Abdomen	57.60	94.78	34.79	70.73

of 37.6 ns in the indoor environment for the LoS case. Also, considering the maximum excess delay, forehead is again the best position with values of 21.7 ns in indoor environment for the LoS case, and with values of 135 and 152 ns in indoor and outdoor environments of the NLoS case. However, the arm and waist were better locations in outdoor environment for the LoS case with maximum excess delay of 12.7 ns each.

Finally, examining the path loss closely and delay for different postures, it can be safely concluded that both positions have relatively the same path loss and delay while in some postures, delay for one posture is better than that for the other but path loss is quite the opposite. Overall, the best position is the forehead given the combination of low path loss and low delay for UWB wireless tag irrespective of the environment, cases (NLoS or LoS), and different body postures.

3.4 Statistical Analysis

Till now, we analyzed the path loss and time dispersion properties of UAV2W systems in different environments and for different body locations. Now, we will look into the modeling of UWB channel fading. For that, in our study, we have used the second order[3] Akaike (AIC) [22, 23] to determine the best fit distribution for the UWB fading radio channel path loss collected during the measurement campaign. A lower AIC value represents the best fit. Mathematically, the second-order AIC is given as

$$\text{AIC}_c = -2\ln L + 2k + \frac{2k(k+1)}{n-k-1},\tag{3.8}$$

where L is the maximized likelihood score, k is the total number of estimated parameters, and n is the total sample size. The relative corrected AIC is given as

$$\Delta = \text{AIC}_c - \min(\text{AIC}_c),$$

which implies that zero value will indicate the best fit. The goodness of fit was compared with Normal, Weibull, Lognormal, Rayleigh, Nakagami, Rician, Gamma, and Exponential distributions.

Based on the results (Figure 3.9), lognormal distribution was the best fit. A sample result is shown in Table 3.9. Using the estimated lognormal

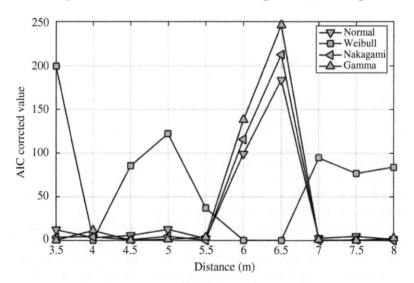

Figure 3.9 Statistical test (AIC) to determine the best distribution for fading between a UAV and human subject at several distance points.

3 Second order was chosen because the path loss is less and also the distribution used to check goodness of fit does not have similar number of parameters.

Table 3.9 AIC score for all the distributions considered for modeling the fading characteristics.

Position	Normal	Weibull	Lognormal	Rayleigh	Nakagami	Rician	Gamma	Exponential
Forehead	0.0036	42.19	17.27	4113.59	1.95	0	8.47	6292.23
Heart	3.49	38.6	4.22	4010.94	0	3.44	1.42	6186.814
Wrist	15.65	252.95	0	5668.19	8.85	15.62	3.75	7872.35
Abdomen	0	19.17	24.12	4513.76	5.12	0.021	13.75	6701.5
Thigh	41.39	420.11	0	5807.27	26.17	41.34	12.42	8012.53
Arm	0.0079	44.176	10.10	4537.45	0.575	0	4.6	6725.8
Shin	0	61.33	17.31	5166.476	3.71	0.01	9.75	7365.04
Waist	16.28	294.24	0	6103.48	9.62	16.26	4.25	8311.30
Ear	11.11	301.88	0	6380.04	6.31	11.10	2.59	8589.74

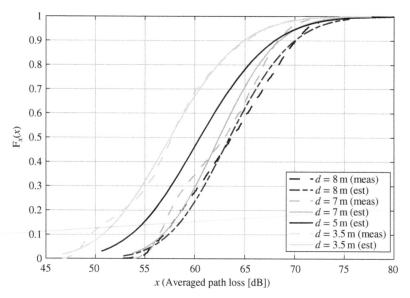

Figure 3.10 Empirical and predicted CDF for radio channel between forehead and UAV in the indoor environment at diagonal distances of 3.5, 5, and 8 m.

distribution, the cumulative distribution fitting (CDF) at two distances of 5.5 and 7.5 m is compared with the empirical CDF, which shows that the estimate is very close to the empirical one (Figure 3.10).

3.5 Conclusion

In this chapter, an in-depth analysis characterizing off-body UWB channel between a human subject and a UAV at 7.5 GHz bandwidth has been done. The transmitter antenna was placed on a UAV while the receiver antenna was patched on a human subject at different body locations during the campaign. The measurement campaign was done in indoor and outdoor environments with LoS and NLoS cases. It was found that the UWB fading channel follows a lognormal distribution from the Akaike (AIC) second-order fitness test. In addition, the different tag locations were analyzed for path loss and delay, and it was concluded that the forehead was the best location to place the UWB antenna in the indoor or outdoor environments for both the cases (LoS and NLoS). Also, considering the different postures with the best combination of path loss and delay, the forehead was again found to be the best location to place the UWB tag.

Bibliography

1 Kachroo, A., Vishwakarma, S., Dixon, J.N. et al. (2019). Unmanned aerial vehicle-to-wearables (UAV2W) indoor radio propagation channel measurements and modeling. *IEEE Access* 7: 73741–73750.

2 Khan, M.M., Abbasi, Q.H., Alomainy, A., and Hao, Y. (2012). Performance of ultrawideband wireless tags for on-body radio channel characterisation. *International Journal of Antennas and Propagation* 2012: 10.

3 Khan, M.M., Abbasi, Q.H., Alomainy, A. et al. (2013). Experimental characterisation of ultra-wideband off-body radio channels considering antenna effects. *IET Microwaves, Antennas & Propagation* 7 (5): 370–380.

4 Hall, P.S. and Hao, Y. (2006). Antennas and propagation for body centric communications. *EuCAP 2006. First European Conference on Antennas and Propagation, 2006*, IEEE, pp. 1–7.

5 Foerster, J., Green, E., Somayazulu, S., Leeper, D. et al. (2001). Ultra-wideband technology for short-or medium-range wireless communications. *Intel Technology Journal* 2: 2001.

6 Allen, B., Dohler, M., Okon, E. et al. (2007). *UWB Antenna and Propagation for Communications, Radar and Imaging.* Wiley: Hoboken, NJ.

7 Schootman, M., Nelson, E.J., Werner, K. et al. (2016). Emerging technologies to measure neighborhood conditions in public health: implications for interventions and next steps. *International Journal of Health Geographics* 15 (1): 20.

8 Lum, M.J.H., Rosen, J., King, H. et al. (2007). Telesurgery via unmanned aerial vehicle (UAV) with a field deployable surgical robot. *Studies in Health Technology and Informatics* 125: 313–315.

9 Todd, C., Watfa, M., El Mouden, Y. et al. (2015). A proposed UAV for indoor patient care. *Technology and Health Care* 1–8 https://europepmc.org/article/med/26409533.

10 Fleck, M. (2016). Usability of lightweight defibrillators for UAV delivery. *Proceedings of the 2016 CHI Conference Extended Abstracts on Human Factors in Computing Systems*, ACM, pp. 3056–3061.

11 Tatham, P., Stadler, F., Murray, A., and Shaban, R.Z. (2017). Flying maggots: a smart logistic solution to an enduring medical challenge. *Journal of Humanitarian Logistics and Supply Chain Management* 7 (2): 172–193.

12 Patrick, W.G. (2016). Request apparatus for delivery of medical support implement by UAV. April 5US Patent 9,307,383.

13 Hu, Z.H., Nechayev, Y.I., Hall, P.S. et al. (2007). Measurements and statistical analysis of on-body channel fading at 2.45 GHz. *IEEE Antennas and Wireless Propagation Letters* 6: 612–615.

14 Alomainy, A., Hao, Y., Owadally, A. et al. (2007). Statistical analysis and performance evaluation for on-body radio propagation with microstrip patch antennas. *IEEE Transactions on Antennas and Propagation* 55 (1): 245–248.

15 Nechayev, Y.I., Hu, Z.H., and Hall, P.S. (2009). Short-term and long-term fading of on-body transmission channels at 2.45 GHz. *Antennas & Propagation Conference, 2009. LAPC 2009. Loughborough*, IEEE, pp. 657–660.

16 Abbasi, Q.H., Sani, A., Alomainy, A., and Hao, Y. (2009). Arm movements effect on ultra wide-band on-body propagation channels and radio systems. *Antennas & Propagation Conference, 2009. LAPC 2009. Loughborough*, IEEE, pp. 261–264.

17 Abbasi, Q.H., Sani, A., Alomainy, A., and Hao, Y. (2010). On-body radio channel characterization and system-level modeling for multiband OFDM ultra-wideband body-centric wireless network. *IEEE Transactions on Microwave Theory and Techniques* 58 (12): 3485–3492.

18 Alomainy, A., Abbasi, Q.H., Sani, A., and Hao, Y. (2009). System-level modelling of optimal ultra wide-band body-centric wireless network.

Microwave Conference, 2009. APMC 2009. Asia Pacific, IEEE, pp. 2188–2191.

19 3DR drones. https://3dr.com/products/supported-drones/ (accessed 10 March 2021).

20 Goldsmith, A. (2005). *Wireless Communications*. Cambridge University Press.

21 Rappaport, T.S. (1996). *Wireless Communications: Principles and Practice*, vol. 2. Hoboken, NJ: Prentice Hall PTR

22 Fort, A., Desset, C., De Doncker, P. et al. (2006). An ultra-wideband body area propagation channel model-from statistics to implementation. *IEEE Transactions on Microwave Theory and Techniques* 54 (4): 1820–1826.

23 Burnham, K.P. and Anderson, D.R. (2003). *Model Selection and Multimodel Inference: A Practical Information-Theoretic Approach*. Springer Science & Business Media.

4

A Cooperative Multiagent Approach for Optimal Drone Deployment Using Reinforcement Learning

Rigoberto Acosta-González[1], Paulo V. Klaine[2],
Samuel Montejo-Sánchez[3], Richard D. Souza[4], Lei Zhang[2], and
Muhammad A. Imran[2]

[1] *Department of Electronics and Telecommunications, Universidad Central "Marta Abreu"*
de Las Villas, Santa Clara, Cuba
[2] *Electronics and Nanoscale Engineering Department, University of Glasgow, Glasgow, UK*
[3] *Programa Institucional de Fomento a la I+D+i, Universidad Tecnológica Metropolitana,*
Santiago, Chile
[4] *Department of Electrical and Electronics Engineering, Federal University of Santa Catarina,*
Florianóplis, Brazil

Acronyms

BS	base station
DDPG	deep deterministic policy gradient
DQN	deep Q-network
DRL	deep reinforcement learning
DSCs	drone small cells
EIRP	equivalent isotropically radiated power
HD	high-density users
ITU-R	International Telecommunication Union-Radio
LD	low-density users
LoS	line-of-sight
LTE	long-term evolution
MARL	multiagent reinforcement learning
MD	medium-density users

Autonomous Airborne Wireless Networks, First Edition.
Edited by Muhammad Ali Imran, Oluwakayode Onireti, Shuja Ansari, and Qammer H. Abbasi.
© 2021 John Wiley & Sons Ltd. Published 2021 by John Wiley & Sons Ltd.

Acronyms

NLoS	non-line-of-sight
PL	path loss
RAN	radio access network
RB	resource block
RL	reinforcement learning
RSRP	received signal power
SINR	signal to interference plus noise ratio
UAV	unmanned aerial vehicle

4.1 Introduction

The past few years have witnessed a major revolution in the area of unmanned aerial vehicles (UAVs), commonly known as drones, due to significant technological advances across various fields, ranging from embedded systems to autonomy, control, security, and communications [1]. Drone small cells (DSCs) are aerial devices equipped with communication apparatus to provide support to a wireless network as mobile base stations (BSs) [2]. However, despite many recent advances, the 3D placement of UAVs (independently of their roles) is still a major design challenge in UAV-enabled wireless communication systems [1]. In general, optimizing UAV deployment is challenging due to the fact that it is a function of various parameters, such as UAV channel gain, interference between UAVs, deployment environment, and user mobility [1].

Despite the massive growth in the integration of UAVs in cellular networks, most of the literature still focuses on analytical solutions for positioning UAVs in mobile networks. Although these solutions have their merits, they often require unrealistic assumptions, such as knowledge of the number of users in the network, their positions, or that users remain static. These assumptions are bold and might not be true in most real scenarios, making the proposed solutions ineffective to be used in practical situations, or very specific to certain conditions [3].

In contrast to analytical solutions, other researchers have relied on the application of intelligent techniques to solve UAV-related problems. Regarding these techniques, a particular family of algorithms that has gained a lot of attention in the realm of UAV-based communications are the ones based on reinforcement learning (RL) [4]. When the environment is dynamic and it is challenging to find a fixed or static solution, RL is a great candidate, as its algorithms are able to learn online from experiences and find optimal solutions [5].

In [4] a Q-learning positioning approach is proposed in an effort to find the best 3D placement of multiple DSCs in a scenario where the conventional terrestrial communication infrastructure is not operational or accessible due to a large-scale natural disaster. The goal of this approach was to facilitate the construction of an efficient emergency communication network by maximizing the total network radio coverage while attaining robustness against dynamic network conditions, mobility issues, and interference. On the other hand, Ghanavi et al. [6] propose the optimal placement of a single aerial-BS, based on Q-learning. De Paula Parisotto et al. [7], extends the solution of [4] and propose an intelligent method based on RL to determine the best transmit power allocation and 3D positioning of multiple DSCs in an emergency scenario. The main goal is to maximize the number of users covered by the drones, while considering user mobility and radio access network (RAN) constraints.

Abeywickrama et al. [8] propose RL and deep reinforcement learning (DRL) based methods to deploy UAV-BSs under energy constraints to provide efficient and fair coverage to the ground users, while minimizing inter-UAV collisions and interference to ground users. For the most part, progress in DRL has been focused on settings where a single agent needs to solve an otherwise static task, corresponding to a single-agent setting [9]. One of the greatest challenges of multiagent reinforcement learning (MARL) is the question of credit assignment: since all agents are exploring and learning at the same time, it is difficult for any given agent to estimate the impact of their actions on the overall return [9]. In the context of UAV communications, the authors in [10] develop an MARL framework in which each agent discovers its best strategy according to its local observations. More specifically, Cui et al. [10] propose an agent-independent method, for which all agents conduct a decision algorithm independently but share a common structure based on Q-learning. On the other hand, Huang et al. [11] carefully design a deep Q-network (DQN) for optimizing the UAV navigation by selecting the optimal policy. The DQN is trained so that the agent is capable of making decisions based on the received signal strengths for navigating the UAVs. Lastly, the authors of [12] propose a decentralized DRL based framework to control each UAV in a distributed manner. They design a fully distributed control solution in which a group of UAVs is assigned to fly around a target area in order to provide long-term communication coverage for the ground mobile users. Table 4.1 shows a comparison between the above state-of-the-art solutions in the context of UAV communications using RL methods.

Table 4.1 State-of-the-art UAV positioning solutions using RL.

References	Short description	RL solution
Ghanavi et al. [6]	Proposes the optimal placement of a single aerial-BS that increases the QoS of wireless networks	Q-learning
Klaine et al. [4]	Deploys DSCs to facilitate the construction of an efficient emergency communication network by maximizing the total network radio coverage	Q-learning
De Paula Parisotto et al. [7]	Proposes an intelligent solution to determine the best transmit power allocation and 3D positioning of multiple DSCs in an emergency scenario	Q-learning
Abeywickrama et al. [8]	Deploys UAV-BSs to provide efficient and fair coverage to ground users	Deep Q-learning
Cui et al. [10]	Proposes a dynamic resource allocation scheme for multiple UAVs-enabled communication networks	Multiagent Q-learning
Huang et al. [11]	Optimizes UAV navigation by selecting the optimal policy	Deep Q-network
Liu et al. [12]	Proposes a framework to control each UAV in a distributed manner to provide coverage to ground mobile users	Multiagent deep deterministic policy gradient
Reis et al. [2]	Proposes an algorithm to optimize the positioning and the transmit power of DSCs considering the energy efficiency	Q-learning

Source: Ghanavi et al. [6], Klaine et al. [4], De Paula Parisotto et al. [7].

In this chapter,

- we study how to optimize the deployment of UAVs to establish an aerial cellular network using RL algorithms;
- we implement a Q-learning algorithm, using a cooperative multiagent approach, in which a set of drones must intelligently find their best 3D

positions, the optimal distribution of frequencies, and whether it should be shut down or not. The objective of the proposed solution is to minimize the number of users in outage in the network.

4.2 System Model

Before discussing the proposed MARL solution in detail, we introduce the system model utilized in this chapter. In the system model we will introduce the parameters of the urban environment model. In addition, in the communication model we will analyze how to calculate the loss of path, throughput, and the process of connection between the UAV and users.

4.2.1 Urban Model

The urban environment model follows the International Telecommunication Union-Radio (ITU-R) document [13], which recommends a standardized model for an urban environment with the following parameters:

- α: The ratio of built-up land area to the total land area.
- β: The average number of buildings per square kilometer.
- γ: A scale parameter for the heights of the buildings.

In order to model an urban scenario, the ITU-R statistical parameters and a building distribution using a Manhattan grid are considered, as in [14]. We will use three different distributions of user equipment (UE), which are detailed in Section 4.4.3. Buildings have a square structure of width W and are separated by a distance S as shown in Figure 4.1. The height of the buildings is generated using a Rayleigh distribution with the γ parameter. The selected urban scenario is displayed in a square area of dimensions $L \times L$, while W and S are obtained in Eqs. (4.1) and (4.2), in meters, by Al-hourani et al. [14]:

$$W = 1000\sqrt{\frac{\alpha}{\beta}}, \tag{4.1}$$

$$S = \frac{1000}{\sqrt{\beta}} - W. \tag{4.2}$$

4.2.2 Communications Model

In order to provide a wide coverage and minimize the users in outage, a total of N_d drones are deployed. Each drone is considered to have a dedicated backhaul link, using a microwave link, which is capable of connecting to

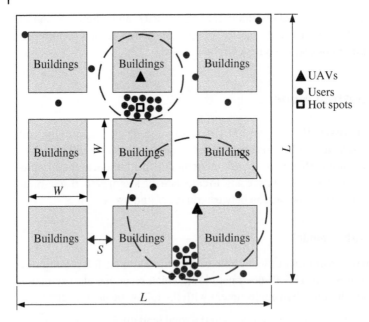

Figure 4.1 Manhattan grid urban layout.

the network operator [4]. Each drone is equipped with an antenna that has a certain directivity. The antenna has a θ opening angle, where the major lobe of the antenna is concentrated and the highest gain is obtained [4]. So, the coverage radius (ρ) of a drone can be calculated as in Eq. (4.3)

$$\rho = h_d \cdot \tan\left(\frac{\theta}{2}\right), \tag{4.3}$$

where h_d is the height of the drone. We also consider that the signal outside the coverage radius is strongly attenuated [4].

The users N_u are distributed in a square area of $L \times L$. A portion of these users are distributed near hot spots around the area and the remaining users are randomly positioned. In Section 4.4.3, we will detail the EU distributions. Let us consider a set $\mathbb{D} = \{1, 2, \ldots, N_d\}$ of drones and a set $\mathbb{U} = \{1, 2, \ldots, N_u\}$ of users. Then, the path loss (PL) between a drone $j \in \mathbb{D}$ and a user $i \in \mathbb{U}$ can be calculated as [15]

$$PL_{ij} = \frac{\Omega}{1 + a\exp(-b[\arctan(\frac{h_d}{R}) - a])} + 10\log(h_d^2 + R^2) + \Psi, \tag{4.4}$$

$$\Omega = \xi_{\text{LoS}} - \xi_{\text{NLoS}} ,$$

$$\Psi = 20\log(f_c) + 20\log(4\pi/c) + \xi_{\text{NLoS}} ,$$

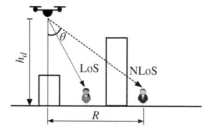

Figure 4.2 UAV path loss in urban environment. Source: Based on Al-Hourani et al. [15].

where R is the Euclidean distance in the horizontal plane between the drone and the user as illustrated in Figure 4.2. The value f_c is the transmission frequency. Three different sets of available frequencies are used with the following distribution: 1 GHz, 1, 1.2, 1.4 GHz, and 1, 1.2, 1.4, 1.6, 1.8, 2 GHz. Note that, all used frequencies are into Sub-6 GHz bands. The coefficients ξ_{LoS} and ξ_{NLoS} are related to the existence of a direct line of sight between the drone and the user, whereas the values of the coefficients a and b can be calculated as [15]

$$z = \sum_{j=0}^{3} \sum_{i=0}^{3-j} C_{ij}(\alpha\beta)^i \gamma^j , \tag{4.5}$$

where the parameter C_{ij} can be obtained from Table I (coefficient a) and Table II (coefficient b) in [15].

The connection of a user i to a drone j depends on the received signal power (RSRP), expressed in dB, and calculated as

$$RSRP_{ij} = EIRP_j - PL_{ij} , \tag{4.6}$$

where $EIRP_{ij}$ is the equivalent isotropically radiated power, and represents the transmitted power combined with the antenna gain of the UAV, in dB [4]. Based on this, the signal to interference plus noise ratio (SINR) can be calculated (in linear scale) as

$$SINR_{ij} = \frac{RSRP_{ij}}{N + \sum_{k=1,k\neq j}^{N_f} RSRP_{i,k}} , \tag{4.7}$$

where N is the additive White Gaussian noise, N_f is the number of drones that have the same transmission frequency as UAV j, and $\sum_{k=1,k\neq j}^{N_f} RSRP_{i,k}$ represents the interference from other UAVs in UAV j.

The throughput T_{ij} of a user i connected to a drone j, in bits per second, can be calculated using the Shannon channel capacity formula [16]

$$T_{ij} = B\log_2(1 + SINR_{ij}), \tag{4.8}$$

where B is the bandwidth in Hz.

The DSCs are considered to have limited resources in both RAN and backhaul. For each drone it is assumed that it shares a 10 MHz bandwidth, which corresponds to a capacity of 50 resource blocks (RBs), according to the long-term evolution (LTE) parameters [4]. Moreover, the DSCs are assumed to have a stable microwave link connection to a central entity.

The process of connecting users to the drones is divided into two steps:

I. The user checks if he/she can stay connected to the drone, verifying if the value of *SINR* is greater or equal to the connection threshold (search threshold in Table 4.2). Otherwise, the user is disconnected.

II. If the SINR of a user is above a certain threshold and the BS has enough space in its RAN, then a user is allocated to that BS in that time slot. However, if that BS has no RBs available or the user SINR is too low (below the connection threshold), the next BSs are tried, in order of the highest SINR. After all BSs are tried, if a user is still unable to be associated with a BS, the user is considered to be out of coverage (in outage) for that time slot [4].

4.3 Reinforcement Learning Solution

Markov decision processes (MDPs) are a tool for modeling sequential decision-making problems where a decision maker interacts with a system in a sequential fashion [21]. MDPs consist of states, actions, transitions between states, and a reward function. An MDP is a tuple $\langle S, A, T, R \rangle$ in which S is a finite set of states, A is a finite set of actions, T is a transition function defined as $T : S \times A \times S \rightarrow [0, 1]$ ($\sum_{s' \in S} T(s, a, s') = 1$), and R is a reward function defined as $R : S \times A \times S \rightarrow R$ [22].

In order to discuss the order in which actions occur, we will define a discrete *global clock*, t. The system being controlled is Markovian if the result of an action does not depend on the previous actions and visited states (history), but only depends on the current state [22]:

$$P(s_{t+1} = s' | a_t = a, s_t = s) = T(s, a, s'). \tag{4.9}$$

4.3.1 Fully Cooperative Markov Games

Markov games or stochastic games are the foundation for much of the research in MARL. Markov games are a superset of MDPs and matrix games, including both multiple agents and multiple states [23].

Table 4.2 Simulation parameters.

Parameters	Value	References
Ratio of built-up to total land area (α)	0.3	ITU-R [13]
Average number of buildings (β)	500	ITU-R [13]
Scale parameter for building heights (γ)	15 m	ITU-R [13]
ξ LoS	1 dB	Al-hourani et al. [14]
ξ NLoS	20 dB	Al-hourani et al. [14]
Side of the square area (L)	1 km	
Drone x-axis and y-axis step	50 m	
Drone z-axis step	100 m	
Minimum drone height	200 m	
Maximum drone height	1000 m	
Number of users (N_u)	768	Jaber et al. [17, 18]
Number of hot spots	16	
Number of DSCs	20	
Ratio of users in near hot spots	2/3	Jaber et al. [17, 18]
DSC EIRP	−3 dBW	Azari et al. [19]
DSC antenna directivity angle (θ)	60°	Full Band 4G Antennas [20]
RBs in DSC	50	Jaber et al. [17, 18]
Bandwidth of one RB	180 kHz	Jaber et al. [17, 18]
Threshold SINR requirement	−3 dB	
Carrier frequency f_c	$[1:0.2:2]^{a)}$GHz	
Total number of episodes	100	
Number of independent runs	100	
Maximum iterations per episode (IT_{max})	1000	
Maximum iterations, same reward (ITR_{max})	100	
Maximum reward (R_{max})	768	
Learning rate (λ)	0.9	Klaine et al. [4]
Discount factor (ϕ)	0.9	Klaine et al. [4]

a) The frequencies used correspond to the set {1, 1.2, 1.4, 1.6, 1.8, 2}GHz in Sub-6 GHz.
Source: Al-hourani et al. [14], Azari et al. [19], Klaine et al. [4].

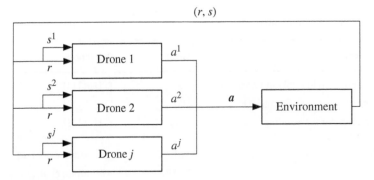

Figure 4.3 MARL framework for multi-drone networks. Source: Based on Cui et al. [10].

A Markov game [24] is defined as a tuple $\langle j, S, A^1, \ldots, A^j, T, R^1, \ldots, R^j \rangle$ where

- j is the number of agents;
- S is the finite set of states;
- A^j is the set of actions available to the agent j (and $\mathbf{a} = a^1 \times \cdots \times a^j$ the joint action set, as shown in Figure 4.3);
- $T : S \times \mathbf{A} \times S \rightarrow [0; 1]$ is the transition function such that $\forall s \in S, \forall \mathbf{a} \in \mathbf{A}, \sum_{s' \in S} T(s, \mathbf{a}, s') = 1$;
- $R_i : S \times \mathbf{A} \rightarrow R$ is the reward function for agent i.

Figure 4.3 describes how a multiagent RL framework can work, in which agents collect information locally about the environment, given by the current state and reward at time step t, and independently decide for their own actions (a^n) based on their own Q-tables. After that, the actions are combined into a single joint action, which is then evaluated at the environment, yielding new states and rewards and repeating the cycle.

The transition and reward functions depend on the joint action. The transition function T gives the probability that action $\mathbf{a}$ in state s at time step t will lead to state s' at step $t + 1$ [24]:

$$P(s_{t+1} = s' | \mathbf{a}_t = \mathbf{a}, s_t = s) = T(s, \mathbf{a}, s'). \tag{4.10}$$

It is assumed that the transition and reward functions are unknown to the agent, but these are discovered by interacting with the environment [9]. We use *cooperative* to refer to the configuration in which all agents get the same reward, elaborating a team reward [9]. When the game is stochastic, the problem is to distinguish between different sources that cause the variation in the observed rewards. The variation can be due to the noise in the environment or to the behavior of the other agents [23].

4.3.2 Decentralized Q-Learning

Q-learning is probably the most used algorithm in the single-agent framework because of its simplicity and robustness [23], and it was also one of the first RL algorithms applied to multiagent environments [25]. Nowadays, several variants of the Q-learning algorithm have been proposed for multiagent environments. In decentralized Q-learning, no coordination problems are explicitly addressed. However, it has been applied with success in some applications [23].

In decentralized Q-learning the Q^j-table of agent j defines the value of a state, s_t, at the time t, to select the a_t^j action and r_{t+1} is the received reward [23]. The update equation for agent j is

$$Q^j(s_t, a_t^j) \leftarrow Q^j(s_t, a_t^j) + \lambda(r_{t+1} + \phi \max_a Q^j(s_{t+1}, a) - Q(s_t, a_t^j)), \quad (4.11)$$

where λ is the learning rate, dictating how fast an agent learns, and ϕ is the discount factor.

The following are the parameters for the decentralized Q-learning algorithm:

- *Agents*: Each drone is an independent agent and has an individual Q-table.
- *States*: States are made up of three components, the 3D position of the drone, its transmission state (a binary value indicating if the drone is "on" or "off"), plus the transmission frequency ($s = [x, y, z, \text{status}_{tx}, f_{tx}]$).
- *Actions*: Each drone can select one action from the set of available actions, the action space. Note that due to its importance and complexity, the detailed description of the action selection process is addressed independently in Section 4.3.3.
- *Reward*: The total number of covered users,

$$R = \sum_{j=1}^{N_d} |U_j|, \quad (4.12)$$

where U_j is the set of users connected to the drone j and $|U_j|$ is the number of users in the set U_j. It is assumed that the drones have access to the total number of users connected to other UAVs through a central entity. Note that the drones must have a backhaul connection to the core network, and that the core entity is on the core network. The backhaul is assumed to be ideal [7].

- *Policy*: Each drone selects an action according to an ϵ-greedy policy [5], which can be expressed as

$$a^j = \begin{cases} \tau, & \text{if rand}(0,1) < \epsilon, \\ \arg\max_{a \in A} Q(s_{t+1}, a), & \text{otherwise,} \end{cases} \qquad (4.13)$$

where τ is a randomly selected action, and rand(0,1) denotes a random number using a uniform distribution over the interval $[0, 1]$.
- *Update*: Each drone updates its Q-table using (4.11).
- *Start*: All drones are initialized in random positions, the Q-table is initialized with all entries set to zero, and the number of available transmission frequencies is established using $\mathbb{F}$.
- *Episode*: Agent–environment interactions are divided into episodes. In this chapter users do not move between episodes. An episode can be described as a snapshot of the network. In each episode, the drones take measurements based on their current state and evaluate their reward. This process is repeated over a number of iterations until one of the stop criteria is met. At the end of an episode, the drones move to the best reward state.
- *Stop criteria*: Three stop criteria are proposed:
  1. The drones have reached the maximum number of iterations for that episode IT_{max}.
  2. The reward has not improved after a number of iterations ITR_{max}.
  3. The reward has reached its maximum value R_{max}. The drones have associated the maximum number of users.

The proposed method is summarized in Algorithm 4.1.

4.3.3 Selection of Actions

Applying RL to practical control tasks requires intelligent and effective learning algorithms, since long learning times are needed to train the algorithms to solve complex problems [26].

In line 7 of Algorithm 4.1 each drone must choose an action using the ϵ-greedy policy. This policy selects either a random action or a learned action.

In order to optimize the learning process, four different strategies were implemented when the agent selects an action. Unlike in the ϵ-greedy policy, in which a random action is chosen to τ, the four different strategies consist of changing how τ is selected. The strategies are the following:

- **Basic strategy**: The action space is as follows: up, down, left, right, forward, backward, stop, turn on, turn off, and change frequency, as shown

---

Algorithm 4.1: Positioning and Frequency Distribution

---

1 Initialise position of the drones
2 Initialize Q-tables
3 Select the number of available transmission frequencies
4 **for** *for each episode* **do**
5   **while** *stop criteria does not occur* **do**
6   **for** $j = 1: N_d$ **do**
7   Select an action A^j using ϵ-greedy
8   Update the state of the drones after using **a**
9   Perform the user connection process using (4.7)
10   Get the reward R_{t+1} according to (4.12)
11   **for** $j = 1: N_d$ **do**
12   Update the Q^j table using (4.11)
13   Move the drones to the position of greatest reward
14   Connect users
15   Get metrics

Figure 4.4 Available action sets. (a) Basic strategy action space. (b) All strategy action space. (c) New strategy action space.

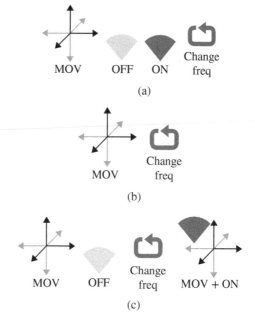

MOV   OFF   ON   Change freq

(a)

Change freq
MOV

(b)

MOV   OFF   Change freq   MOV + ON

(c)

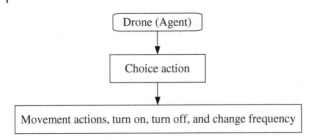

Figure 4.5 **Basic strategy.**.

in Figure 4.4a. All the actions are equiprobable. There is no optimization in the action selection process. Figure 4.5 describes this strategy in the process of action selection.

- **ALL strategy**: The action space is as shown in Figure 4.4b. The drones always have their radio transmitters on. As shown in Figure 4.6, selecting the "frequency change" action or choosing the motion actions depends on the drone capacity. If its capacity is greater than 80%, the drone chooses to move, whereas if the capacity is less than 80% then it changes frequency.
- **New strategy**: The action space is as shown in Figure 4.4c. Multiple or compound actions are used. Selection of an action in this strategy depends on whether the drone is transmitting or not, as shown in Figure 4.7. If the radio transmitter is off, the actions to be chosen are to turn the radio on and carry out a movement; to speed up the learning process and avoid states that oscillate between the ON and OFF of the radio transmitter, without contributing to the search for better positioning or learning. With probability $1 - p$ the drone shuts down and with probability p the drone can choose to move or change frequency. This probability (p) is calculated from the capacity of the UAV as shown in Figure 4.7. Selecting the

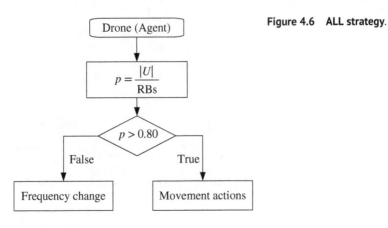

Figure 4.6 **ALL strategy**.

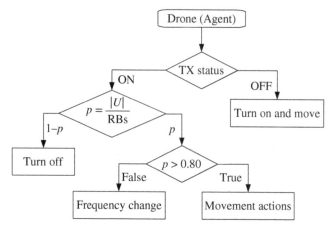

Figure 4.7 **New strategy.**

frequency changing action or the movement actions depends on the current capacity, and is a process similar to the **ALL strategy**.

- **Hybrid strategy**: This method starts with the **ALL strategy** and then employs the **New strategy** when the following criteria is met: the difference in reward between two continuous episodes is less than or equal to 1 for five consecutive episodes. This change of strategy is an irreversible process.

In addition to these strategies, there are also some restrictions. For example, if only one frequency is available, the frequency change action is removed from the action group in all strategies. Therefore, the process of selecting actions in each of the strategies is slightly modified, as it is no longer necessary to know the drone's ability to decide whether to select a motion action or the frequency changing action; thus the drone automatically chooses a motion action. Moreover, it is important to note that the number of actions in each strategy is different. The decision to change the transmission frequency is determined by the agent (drone), so the mechanism is simple. The frequency changing action works as a sequential rotation between the available frequencies.

4.3.4 Metrics

In order to evaluate the strategies, the following metrics are used:

1. The percentage of users in outage, given by

$$D_u = 100 \cdot \frac{N_o}{N_u},$$ (4.14)

where N_o is the number of users in outage and it is calculated as

$$N_o = N_u - \sum_{j=1}^{N_d} |U_j|. \tag{4.15}$$

2. Average global system backhaul throughput (B_{global}) per episode. We get B_{global} as follows:

$$B_{\text{global}} = \frac{\sum_{j=1}^{N_d} \sum_{i=1}^{U_j} T_{i,j}}{F_{\text{all}}}, \tag{4.16}$$

where F_{all} is the number of active transmission frequencies and $T_{i,j}$ is obtained using (4.8).

3. Average number of active drones, which is counted per episode as N_{on}:

$$N_{\text{on}} = \sum_{j=1}^{N_d} \chi_j, \tag{4.17}$$

$$\chi = \begin{cases} 1, & \text{radio transmitter is ON,} \\ 0, & \text{otherwise.} \end{cases} \tag{4.18}$$

4.4 Representative Simulation Results

4.4.1 Simulation Scenarios

To demonstrate the effectiveness of the proposed solution, several simulation scenarios have been implemented in Python. The main simulation parameters are shown in Table 4.2.

4.4.2 Environment

The simulation scenarios consist of an urban area of 1 km^2, complying with the model and parameters established in [14]. In this area 768 users are distributed, and each one of the users can calculate their throughput using (4.8) where the SINR is obtained from (4.7).

4.4.3 User Distribution

In this environment, three different types of user distribution are investigated. Scenarios with low, medium, and high density are generated and the placement of users follows a similar manner, with only the density of users varying between them. For the three distributions, one third of the users are placed at random throughout the entire area, while two thirds are randomly

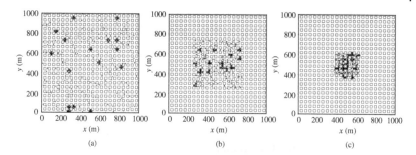

Figure 4.8 User density areas. (a) Low density. (b) Medium density. (c) High density.

assigned close to artificially generated points, with each point containing the same number of users. The densities of each user distribution are implemented as follows:

- *Low density*: Users can be placed anywhere in the area, according to Figure 4.8a.
- *Medium density*: Users are distributed centrally occupying half of the 1 km^2 area, according to Figure 4.8b.
- *High density*: Users are distributed centrally occupying a quarter of the 1 km^2 area, as in Figure 4.8c.

4.4.4 Simulation

To solve the problem using the proposed algorithm, the movement of the drones was discretized in steps of 50 m in the horizontal plane ($x - y$ axis) and 100 m in the vertical plane (z axis).

For the simulation, 10 scenarios were generated for each user distribution. For each scenario, 100 independent runs were made, each with a total of 100 episodes. Before the drones begin their movements and determine their best positions, a certain number of hot-spots must be generated with an equal number of users per hot-spot. In addition, drones must have their initial positions and Q-tables initialized accordingly. Three different scenarios were tested in the simulation (low, medium, and high user density) and in each scenario the performance was evaluated with three different sets of available frequencies: (i) only a single frequency; (ii) three orthogonal frequencies; (iii) six orthogonal frequencies.

After the initialization process is finished, the drones will try to decrease the number of users in outage in each episode. To make this possible each episode is divided into iterations. For each iteration the drones move together through the environment looking for the best positions. When some of the stop criteria are reached the episode is stopped, returned to the best position found, and the metrics are obtained.

4.4.5 Numerical Results

4.4.5.1 Single Frequency

As we can see from Figure 4.9 the **ALL strategy** achieves the best results, and the **New and Hybrid strategies** maintain similar values in their results but inferior to the **ALL strategy**, while the **Basic strategy** achieves the least efficient results for all distributions. The increase in users in outage as the density of user distribution increases causes the Global System Backhaul to decrease as illustrated in Figure 4.10, demonstrating the greater efficiency

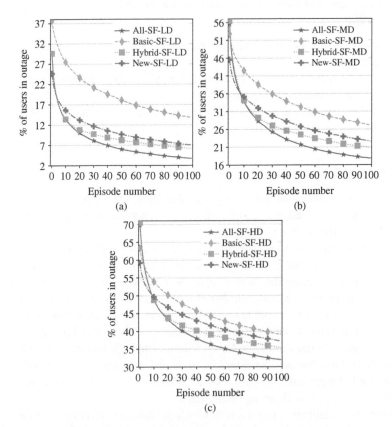

Figure 4.9 **Single frequency:** Number of users in outage. (a) Low density. (b) Medium density. (c) High density.

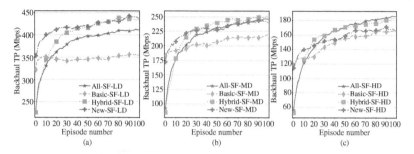

Figure 4.10 **Single frequency:** Global system backhaul. (a) Low density.
(b) Medium density. (c) High density.

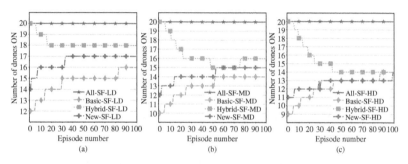

Figure 4.11 **Single frequency:** Number of active drones. (a) Low density.
(b) Medium density. (c) High density.

of the **ALL strategy** when a single frequency is used and the density of users increases. In addition, it seems that the **ALL strategy** comes to know topologies where some drones are located almost out of the area to serve isolated users. As expected with the increase in the density of users, there is a greater benefit in decreasing the number of active drones as illustrated in Figure 4.11. Moreover, the Hybrid approach gives almost as good results as the ALL approach, but with 40% less drones.

4.4.5.2 Three Frequencies

As expected, with the increase in the number of operating frequencies, the number of users in outage decreases significantly, as illustrated in Figure 4.12. As can be seen from Figure 4.12a, when the user density is low, the **New strategy** surpasses the **ALL strategy**, as if the latter strategy was unable to learn the benefit of using different frequencies. So much so that even the **Basic strategy** surpasses the **ALL strategy** when higher user densities are considered.

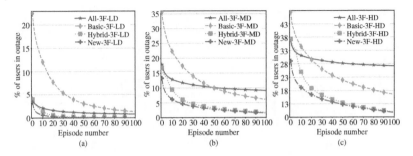

Figure 4.12 Three frequencies: Number of users in outage. (a) Low density. (b) Medium density. (c) High density.

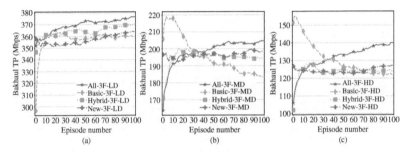

Figure 4.13 Three frequencies: Global system backhaul. (a) Low density. (b) Medium density. (c) High density.

With the increase to three operating frequency bands, the system's performance in terms of users in outage (Figure 4.12b) is superior to the low user density scenario with a single frequency of operation (Figure 4.9a). Although there is an apparent benefit in terms of increased backhaul throughput for the **ALL strategy**, as observed in Figure 4.13, the average backhaul throughput per drone is higher in the other strategies. Note that increasing the operating frequencies increases the convergence of the three strategies that switch the drones to OFF in terms of the optimal number of drones, as reflected in Figure 4.14. Therefore, a greater number of UAVs can coexist without generating interference between them and therefore the schemes with the ability to turn off the UAVs converge in higher values than when a single frequency is available.

4.4.5.3 Six Frequencies

With the increase to six frequency bands, the performance of the system in the high density scenario (Figure 4.15a) is superior in terms of outage to the low density scenario with a single operating frequency (Figure 4.9a) and to

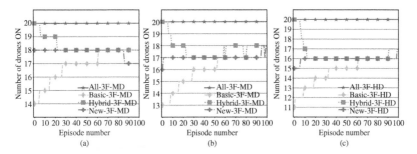

Figure 4.14   Three frequencies: Number of active drones. (a) Low density. (b) Medium density. (c) High density.

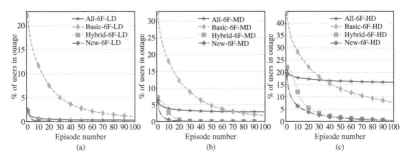

Figure 4.15   Six frequencies: Number of users in outage. (a) Low density. (b) Medium density. (c) High density.

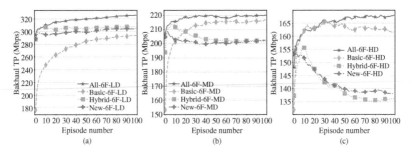

Figure 4.16   Six frequencies: Global system backhaul. (a) Low density. (b) Medium density. (c) High density.

the medium density scenario (Figure 4.12b) when there are three frequencies in operation, as reflected in Figure 4.15. As expected, there is a decrease in the overall backhaul of the system, due to the normalization of the available bandwidth, as shown in Figure 4.16.

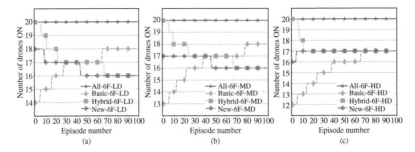

Figure 4.17 **Six frequencies:** Number of active drones. (a) Low density.
(b) Medium density. (c) High density.

Note that the system now finds benefits in keeping more drones active
for the high densities since interference is much smaller when using adjacent drones with different frequencies, as illustrated in Figure 4.17. Note also
that with increasing operating frequencies and scenario density the difference in terms of backhaul throughput between the **ALL strategy** and the
New strategy increases. This indicates that the number of optimal frequencies also depends on the density of users, since Figure 4.17 shows that the
higher density scenario requires one more drone in operation than the low
and medium density scenarios.

4.5 Conclusions and Future Work

4.5.1 Conclusions

From the results obtained in this work we can conclude the following:

- For a single transmission frequency the proposed algorithm obtains better
  performance using the **ALL strategy**, mainly when the density of users
  increases. This is because this strategy tends to locate positions to establish
  coverage of isolated users, which presupposes low energy efficiency and
  sometimes coverage outside the area of interest.
- Using three frequency bands in operation, the **New strategy** outperforms
  the other strategies in terms of users in outage, convergence time, and
  efficient use of resources (in terms of the total number of UAVs).
- By increasing to six operating frequency bands, the **New strategy** shows
  greater effectiveness than the other strategies, finding greater benefit in
  the use of multiple frequencies to avoid interference. However, this fact
  implies a decrease in the efficiency of resources, and consequently, a
  notable decrease in the overall system backhaul throughput.

- As the density of users in the region of interest increases, the required number of operating frequencies increases as well.
- The adoption of multiple actions to the set of actions, as proposed by the **New strategy**, speeds up the convergence of the MA Q-learning algorithm.

4.5.2 Future Work

Based on the developments in of this research, future work should focus on addressing the following:

- The design of reinforcement learning algorithms focused on the cooperation and coordination of UAVs to achieve their goal.
- The design of intelligent strategies for frequency assignment, where optimization of the number of frequencies according to the scenario in question is executed.
- Transmission power control mechanisms, considering the use of beamforming techniques to control the opening angle of the coverage cone.

Acknowledgments

This work has been partially supported in Chile by ANID FONDECYT Iniciación No. 11200659, SCC-PIDi-UTEM FONDEQUIP-EQM180180, and Collaborative Research Activities between PIDi/UTEM and FIE/UCLV, as well as in Brazil by CNPq, and the DARE project grant (No. EP/P028764/1) under the EPSRC's Global Challenges Research Fund (GCRF) allocation.

Bibliography

1 Saad, W., Bennis, M., Mozaffari, M., and Lin, X. (2020). *Wireless Communications and Networking for Unmanned Aerial Vehicles*. Cambridge University Press. ISBN: 9781108691017. https://doi.org/10.1017/9781108691017. https://www.cambridge.org/core/product/identifier/9781108691017/type/book.

2 Reis, A.F., Brante, G., Parisotto, R. et al. (2020). Energy efficiency analysis of drone small cells positioning based on reinforcement learning. *Internet Technology Letters* 5 (5) 4–9. https://doi.org/10.1002/itl2.166.

3 Imran, M.A., Abdulrahman Sambo, Y., Abbasi, Q.H. et al. (2019). Intelligent positioning of UAVs for future cellular networks. *Enabling 5G*

Communication Systems to Support Vertical Industries, pp. 217–232. https://doi.org/10.1002/9781119515579.ch10.

4 Klaine, P.V., Nadas, J.P.B., Souza, R.D., and Imran, M.A. (2018). Distributed drone base station positioning for emergency cellular networks using reinforcement learning. *Cognitive Computation* 10 (5): 790–804. https://doi.org/10.1007/s12559-018-9559-8.

5 Sutton, R.S. and Barto, A.G. (2018). *Reinforcement Learning: An Introduction*. MIT press.

6 Ghanavi, R., Kalantari, E., Sabbaghian, M. et al. (2018). Efficient 3D aerial base station placement considering users mobility by reinforcement learning. *IEEE Wireless Communications and Networking Conference, WCNC*, Volume 2018-April, pp. 1–6. ISBN: 9781538617342. https://doi.org/10.1109/WCNC.2018.8377340.

7 De Paula Parisotto, R., Klaine, P.V., Nadas, J.P.B. et al. (2019). Drone base station positioning and power allocation using reinforcement learning. *Proceedings of the International Symposium on Wireless Communication Systems*, Volume 2019-Augus, pp. 213–217. ISBN: 9781728125275. https://doi.org/10.1109/ISWCS.2019.8877247.

8 Abeywickrama, H.V., He, Y., Dutkiewicz, E. et al. (2020). A reinforcement learning approach for fair user coverage using UAV mounted base stations under energy constraints. *IEEE Open Journal of Vehicular Technology* 1: 67–81. https://doi.org/10.1109/ojvt.2020.2971594.

9 Foerster, J.N. (2018). Deep multi-agent reinforcement learning. PhD thesis. University of Oxford. https://ora.ox.ac.uk/objects/uuid:a55621b3-53c0-4e1b-ad1c-92438b57ffa4 (accessed 15 March 2021).

10 Cui, J., Liu, Y., and Nallanathan, A. (2020). Multi-agent reinforcement learning-based resource allocation for UAV networks. *IEEE Transactions on Wireless Communications* 19 (2): 729–743. https://doi.org/10.1109/TWC.2019.2935201.

11 Huang, H., Yang, Y., Wang, H. et al. (2020). Deep reinforcement learning for UAV navigation through massive MIMO technique. *IEEE Transactions on Vehicular Technology* 69 (1): 1117–1121. https://doi.org/10.1109/TVT.2019.2952549.

12 Liu, C.H., Ma, X., Gao, X., and Tang, J. (2020). Distributed energy-efficient multi-UAV navigation for long-term communication coverage by deep reinforcement learning. *IEEE Transactions on Mobile Computing* 19 (6): 1274–1285. https://doi.org/10.1109/TMC.2019.2908171.

13 ITU-R (2012). Propagation data and prediction methods required for the design of terrestrial broadband radio access systems in a frequency range from 3 to 60 GHz. *Rec. ITU-R P.1410-5*, pp. 1–18.

14 Al-hourani, A., Kandeepan, S., and Jamalipour, A. (2016). Modeling air-to-ground path loss for low altitude platforms in urban environments. *IEEE Transactions on Vehicular Technology* 66 (1): 632–636. https://doi.org/10.1109/GLOCOM.2014.7037248.

15 Al-Hourani, A., Kandeepan, S., and Lardner, S. (2014). Optimal LAP altitude for maximum coverage. *IEEE Wireless Communications Letters* 3 (6): 569–572. https://doi.org/10.1109/LWC.2014.2342736.

16 Goldsmith, A. (2005). *Wireless Communications*. Cambridge University Press. volume 9780521837. ISBN: 9780511841224. https://doi.org/10.1017/CBO9780511841224.

17 Jaber, M., Imran, M., Tafazolli, R., and Tukmanov, A. (2015). An adaptive backhaul-aware cell range extension approach. *2015 IEEE International Conference on Communication Workshop, ICCW 2015*, number BackNets, pp. 74–79. ISBN: 9781467363051. https://doi.org/10.1109/ICCW.2015.7247158.

18 Jaber, M., Imran, M.A., Tafazolli, R., and Tukmanov, A. (2016). A distributed SON-based user-centric backhaul provisioning scheme. *IEEE Access* 4: 2314–2330. https://doi.org/10.1109/ACCESS.2016.2566958.

19 Azari, M.M., Rosas, F., and Pollin, S. (2018). Reshaping cellular networks for the sky: major factors and feasibility. *IEEE International Conference on Communications*, volume 2018-May. ISBN: 9781538631805. https://doi.org/10.1109/ICC.2018.8422685.

20 Full Band 4G Antennas (2017). Outdoor 4G Antenna - Cross Polarised 3G/4G Antenna. https://www.fullband.co.uk/product/fbxpmimo/ (accessed 15 March 2021).

21 Szepesvári, C. (2010). Algorithms for reinforcement learning. *Synthesis Lectures on Artificial Intelligence and Machine Learning* 4 (1): 1–103. ISSN: 1939-4608.

22 Wiering, M. and Van Otterlo, M. (2012). *Reinforcement Learning*, vol. 12. Springer. ISBN: 364227644X.

23 Matignon, L., Laurent, G.J., and Le Fort-Piat, N. (2012). Independent reinforcement learners in cooperative Markov games: a survey regarding coordination problems. *The Knowledge Engineering Review* 27 (1): 1–31. https://doi.org/10.1017/S0269888912000057.

24 Shapley, L.S. (1953). Stochastic games. *Proceedings of the National Academy of Sciences of the United States of America* 39 (10): 1095–11100. https://doi.org/10.1073/pnas.39.10.1095.

25 Tan, M. (1993). Multi-agent reinforcement learning: independent vs. cooperative agents. In: *Machine Learning Proceedings 1993*, 330–337. Elsevier. https://doi.org/10.1016/b978-1-55860-307-3.50049-6.

26 Nishiyama, R. and Yamada, S. (2016). *Reinforcement Learning with Multiple Actions.* In: *Proceedings of the 3rd International Conference on Intelligent Technologies and Engineering Systems (ICITES2014).* Springer: Cham.

5

SWIPT-PS Enabled Cache-Aided Self-Energized UAV for Cooperative Communication

Tharindu D. Ponnimbaduge Perera[1] and Dushantha Nalin K. Jayakody[1,2]

[1]*Department of Information Technology, School of Computer Science and Robotics, National Research Tomsk Polytechnic University, Tomsk, Russian Federation*
[2]*Centre for Telecommunication Research, School of Engineering, Sri Lanka Technological Campus, Padukka, Sri Lanka*

5.1 Introduction

In the past few years, unmanned aerial vehicles (UAVs), which are commonly known as drones, have witnessed significant growth and operational success in wireless communication networks. UAVs have great potential in facilitating adaptive communication facilities due to their rapid mobility and deployment flexibility, which can be great assets in cooperative communication networks for civil and military services, i.e. disaster management, intelligent transportation systems, surveillance, etc. Therefore, increasing research interests from both academia and industry have been received in UAV communications to identify fruitful techniques and methods to integrate UAVs with the existing communication infrastructures. Especially, UAV-assisted radio access has been identified as one of the key components of the fifth-generation (5G) cellular networks, where UAVs act as aerial base stations (ABS) to construct a flexible network architecture. Furthermore, cooperative UAV communications have been considered in more diverse 5G services, i.e. Internet-of-Things (IoT) and Internet-of-Everything (IoE), such that the requirements of the user's quality-of-experience (QoE) can be better accommodated.

Recent research in open literature has focused on UAV communications in three main directions, i.e. (i) UAV's trajectory optimization to improve system performance [1–3], (ii) UAVs used as aerial relays [4–6], and (iii) UAVs as ABS [7–9]. In [1], the authors have presented an approach for

Autonomous Airborne Wireless Networks, First Edition.
Edited by Muhammad Ali Imran, Oluwakayode Onireti, Shuja Ansari, and Qammer H. Abbasi.

the optimization of the UAV's trajectory while minimizing fuel consumption and mean revisit time, and maximizing mean probability of detection in maritime radar applications. The authors have used Quintic polynomials to generate UAV trajectories with complete and complex solutions while providing minimum inputs. The trajectory of the UAV was optimized along with its transmit power while minimizing the outage probability of the UAV-assisted communication network in [2]. The proposed trajectory scheme outperformed fixed power and circular trajectory schemes close to the exhaustive search minimum outage probability with a difference less than 5%. In [3], the UAV's trajectory was optimized to improve the overall energy efficiency of the proposed communication system. Further, the authors have created a theoretical model on the propulsion energy consumption of the fixed wing UAV as a function of flying speed, direction, and acceleration. In [4], a self-energized UAV-assisted relaying scheme was proposed to maximize the achievable throughput at the destination. The proposed relaying scheme outperformed the existing similar schemes named harvest-then-cooperate, self-energy recycling, and traditional time-switching. In [5], the authors have proposed two UAV-assisted relaying schemes to serve as high-speed moving source taking full advantage of the flexible mobility of the UAV. An energy-efficient cooperative multi-hop relay scheme was proposed for UAV-assisted IoT network in [6]. In [7], the authors have proposed a multi-antenna transceiver design and multi-hop device-to-device (D2D) links for UAV-assisted IoT communication networks. Moreover, the shortest-path-routine algorithm was designed to construct D2D links with the minimum number of communication links to enhance the coverage of the UAV. In general, the analysis of typical cell coverage in a UAV ABS multitier network is more complex compared to conventional 3G/4G communication systems. Therefore, a general converge probability of a UAV ABS multitier network was derived in [8]. In this regard, the authors have assumed the distance from the nearest interfering UAV ABS to a user equipment as a low limit of integration. To fill the research gap in realistic radio and traffic models for UAV ABS deployment planning, a flow-level model for realistically characterizing the UAV ABS performance was given in [9]. Further, the authors have proposed the deep reinforcement learning approach to learn the optimal trajectory of the UAV depending on the data traffic.

With the exponential increase in the usage of wireless devices, energy consumption and CO_2 emissions increase at an alarming rate. These concerns are becoming more critical as the number of devices will exceed more than 50 billions by early 2021. The recent development of the latest paradigm of IoT applications realizes the reliable interconnection between devices, services, and users with or without the intervention of humans. Almost every communication aspect within the IoT applications requires wireless sensors, which are battery-limited devices where energy is the pivotal resource to maintain the expected functionality within the communication network. Sometimes, these devices may not have a continuous energy supply due to the nature of its location in the area of the communication setup. Therefore, communication networks need to consist of not only efficient and effective policies of energy management, but also alternative ways of energy harvesting (EH) from various energy sources.

Simultaneous wireless information and power transfer (SWIPT) radio frequency (RF) EH technique has received significant attention from the researchers as a potential solution to address the energy constraint problem of the wireless devices in IoT networks. SWIPT enables the concurrent transmission of both data and energy with the aid of RF signals, providing an essential solution for the rapid drainage of batteries in wireless devices. However, alternative receiver architecture is required to facilitate both data reception and EH from a single RF signal, which is impossible with the conventional receiver architecture. Therefore, two noble receiver architectures named time-switching (TS) and power-splitting (PS) have been proposed. In TS architecture, full received power of the signal is used either for EH or for information processing within the given amount of time, which is decided upon considering the QoS requirement at the communication node. In PS architecture, the received signal is divided into two parts using a power-splitter for the given splitting ratio, where one part of the signal is used for EH while the remaining part of the signal is used for information processing. Similar to TS architecture, PS ratio needs to be decided by considering the required QoS at the communication node. The quantum of research on SWIPT-enabled communication systems has increased recently due to the advantages it provides, such as SWIPT can improve the energy efficiency while maintaining the spectrum efficiency of the communication networks. For a detailed survey on the research on SWIPT, see [10, 11] and the references therein.

Network congestion is one of the major problems to be considered when designing communication networks for future communication applications. Especially, during peak hours, data traffic of the communication networks is increased due to the finite network resources with respect to the active number of users. Another main reason identified for network congestion is the repetitive request demands for similar content by the active users. One of the promising solutions for this bottleneck is to use caching mechanism to avoid repetitive requests for similar content. Caching contains two phases, namely placement and delivery. In the placement phase, popular contents are stored in distributed caches near the users during the off-peak time intervals, in which network resources are minimally used. In the second phase, also known as delivery phase, if the user-requested contents are located in the caches nearby, the user can be served without the intervention of the core network, minimizing network congestion. Incorporating caching into the communication setup causes reduction in the backhaul's loads and increases the energy efficiency.

A framework combining both the EH and caching named GreenDelivery is proposed in [12] to facilitate efficient content delivery in small cells with the aid of EH. In [13], the authors have proposed an energy-efficient power control scheme for an EH small cell base station equipped with wireless backhaul and local storage. Furthermore, it is assumed that the energy arrivals are Poisson distributed and the caching mechanism is modeled using Zipf's law. A new caching mechanism for the EH-based IoT services is proposed in [14] and the trade-off between energy consumption and caching of such a system is investigated by developing an analytical model. In [15], the authors have investigated the cost-effective planning of heterogeneous vehicular networks consisting of cache-enabled roadside units. A cache-assisted SWIPT-TS enabled cooperative system is investigated in [16], where the authors have investigated the effect of caching on the achievable throughput at the destination. It is clearly noted that except the work represented in [16], all other works considered either harvesting energy using external energy source or having an intrinsic relationship between EH and caching.

The following are the identified technical challenges associated with using UAV in communication networks. First, due to the lack of fixed-line backhaul, backhaul from the infrastructure to the UAV can be a bottleneck for the system performance, limiting the UAV's effectiveness in serving edge users with high-quality content with the requested data rate. Secondly, an adaptive trajectory design is needed to cater to the needs of edge users. Finally, UAVs are also energy-constraint devices with limited battery capacity. To overcome the aforementioned drawbacks, in this chapter we investigate the SWIPT-PS enabled cache-aided decode-and-forward (DF) UAV-assisted relaying system, where multiple users are served by a cache-aided UAV within the given operational time. On the contrary, caching-aided UAVs have been investigated in [17–20] without considering EH capability for either the UAV or the users. In the proposed communication setup, the UAV's communication capabilities are solely powered by the amount of energy harvested via SWIPT-PS. Next, an optimization problem (P1) is formulated to maximize the receiving information rate for each user considering caching capacity, amount of energy harvested, and

QoS constraints of the system. Then, we formulate an optimization problem (P2) to identify the trajectory of the UAV once the optimal parameters are given from the (P1). The first optimization problem (P1) is solved by using Karush–Kuhn-Tucker (KKT) conditions, and optimal closed-form solutions for the PS factor at the UAV and time ratio for time block structure are obtained. The second optimization problem (P2) is solved via an iterative algorithm proposed in this chapter.

5.2 System Model

A self-energized DF UAV-enabled relaying scheme is considered as illustrated in Figure 5.1, which consists of one source, one UAV, and multiple users. Each communication device within the given communication network is equipped with a single antenna and works in half-duplex mode. There is no direct link available between the source and the users due to constraints in coverage, i.e. blockage, limited transmission power, etc. The UAV is exploited as a relay to assist information transmission between the

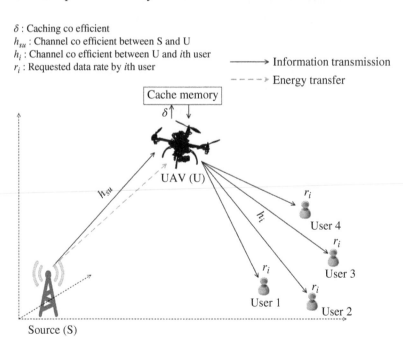

δ : Caching co efficient
h_{su} : Channel co efficient between S and U
h_i : Channel co efficient between U and ith user
r_i : Requested data rate by ith user

⟶ Information transmission
⤏ Energy transfer

Cache memory
δ

UAV (U)
h_{su}
h_i
r_i User 4
r_i User 3
r_i
r_i User 2
User 1

Source (S)

Figure 5.1 Reference system model of self-energized UAV-assisted communication system.

Table 5.1 Definitions of mathematical symbols and variables.

Symbol	Meaning	Symbol	Meaning
ρ	Power-splitting factor	τ	Time allocation for EH
T	Time duration of transmission block	I	Number of users
h_{su}	Channel coefficient between source and UAV	h_i	Channel coefficient between UAV and user
L	Large-scale average channel power gain	l	Small-scale fading coefficient
d_{xy}	Distance between nodes	α	Path loss exponent
$\mathbf{q}[n]$	UAV's location at a given time slot n	$\mathbf{w}[n]$	Location of source and users
K	Rician factor	θ	Elevation angle
g_0	Deterministic LoS component	$\hat{g}_0$	Random scattering component
y_{su}	Received signal at UAV	y_i	Received signal at users
P_s	Transmit power of source	P_u	Transmit power of UAV
n_{su}	AWGN between at UAV	n_i	AWGN at users
σ_{su}^2	Noise variance at UAV	σ_i^2	Noise variance at users
E_u	Energy harvested by the UAV via SWIPT-PS	η	RF to DC energy conversion efficiency
γ_{su}	Received SNR at the UAV	γ_i	Received SNR at the user
B	Channel bandwidth	A_1, A_2	Constant coefficient [21]
R_u	Information rate at UAV	R_i	Information rate at users
δ	Caching coefficient	r_0	User requested information rate
ψ	Ratio of minimum required transmission power of UAV	ϕ_u, ϕ_i	Binary variables for UAV's time scheduling
x_0, y_0	Starting location of UAV	x_d, y_d	Ending location of UAV
V_m	Maximum velocity of UAV	N	Number of time slots

source and the users. The transmission capabilities of the UAV is solely powered by the energy harvested via PS-SWIPT and UAV equipped with a general caching model to store the information received from the source. A block diagram of the proposed system model is given in Figure 5.2, which illustrates the key parameters of the proposed system. All the mathematical symbols used within this chapter are given in Table 5.1.

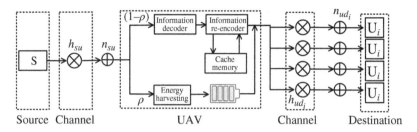

Figure 5.2 Block diagram of the decode-and-forward (DF) relaying for the self-energized UAV.

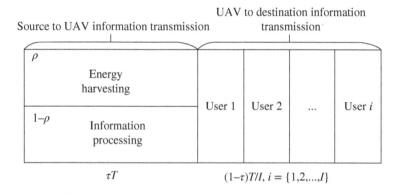

Figure 5.3 Time block diagram of the proposed system model.

Figure 5.3 illustrates the two main phases of the time block diagram of the proposed system. Specifically, in each transmission block of time duration T, the first time fraction τT, where $0 < \tau < 1$, is allocated to information and energy transmission from the source to the UAV. After receiving the signal from the source, the UAV splits it into two parts with the aid of the PS factor ρ, where $0 < \rho < 1$, i.e. one for EH and the rest for information processing. The remaining time fraction $(1 - \tau)T$ is equally divided and allocated for information transmission between the UAV and the users, where $\frac{(1-\tau)T}{I}$ fraction of the time is allocated for each user. It is also noteworthy that the larger T value will increase the achievable throughput at the users due to allocation of more time for information transmission between the UAV and the users. However, it is worth remembering that larger T can also increase the energy consumption and waiting time for each user within the current cycle and up to the next. If there are multiple cycles, $T_{\text{tot}} = \sum_{z=1}^{Z} T$, where Z denotes the number of cycles. Therefore, it is essential to choose the T to maintain a proper balance between the achievable information rate, service delay, and the energy consumption. The period of T is divided into N time slots equal in size as $T = N\zeta$, where ζ is sufficiently small such that the UAV's location does not change considerably within each slot.

5.2.1 Air-to-Ground Channel Model

All the channels are quasi-static[1] and subjected to independent but non-identically distributed Rician fading. The channel fading coefficient between the base station and the UAV is denoted by h_{su} and the channel fading coefficient between the UAV and the users is denoted by h_i, with $i \in \{1, 2, \dots, I\}$. Therefore, the channel between the communication nodes can be modeled as

$$(h_{su}, h_i)[n] = \sqrt{L[n]}l[n], \tag{5.1}$$

where $L(n)$ denotes the large-scale average channel power gain considering both path loss and shadowing, and $l[n]$ denotes the small-scale fading coefficient. Then, $L[n]$ can be written as

$$L[n] = L_0 d_{xy}^{-\alpha}[n], \tag{5.2}$$

where L_0 denotes the average channel power gain at a reference distance, α denotes the path loss exponent and d_{xy}, where $x, y \in \{s, u, i\}$ is the line-of-sight (LoS) distance between the communication nodes. The LoS distance from the UAV $d_{x,y}$ can be written as $d_{x,y} = \sqrt{||\mathbf{q}[n] - \mathbf{w}[n]||^2 + H^2[n]}$, where $\mathbf{w}[n] = (x[n], y[n])$ denotes the location of the source or the users, $\mathbf{q}[n] = (x[n], y[n])$ denotes the UAV's location at the given time block $[n]$, and H denotes the flying altitude of the UAV. Next, considering the LoS path, $l[n]$ can be modeled as

$$l[n] = \sqrt{\frac{K[n]}{K[n] + 1}}g_0 + \sqrt{\frac{1}{K[n] + 1}}\hat{g}_0, \tag{5.3}$$

where g_0 denotes the deterministic LoS component, $\hat{g}_0$ denotes the random scattering component with a zero-mean unit-variance circularly symmetric complex Gaussian (CSCG) [22], and $K[n]$ represents the Rician factor[2] of the channel between UAV and the communication nodes of time block $[n]$. Considering the elevation angle, the Rician factor can be modeled as [21, 22]

$$K[n] = A_1 \exp A_2 \theta[n], \tag{5.4}$$

---

1 The channel fading coefficient remains constant within a transmission block, but may change independently over the next blocks [3].
2 Rician factors in the same time slot are assumed to be identical due to the negligible change in elevation angle in each time slot.

where $\theta[n] = arcsin(H[n]/d_{xy}[n])$ and A_1, A_2 are constant coefficients determined by the specific environment [21].

5.2.2 Signal Structure

The signal received at the UAV and the split signal for information processing can be written as

$$\underbrace{y_{su} = \sqrt{P_s}h_{su}x + n_{su},}_{\text{Received signal at the UAV}} \tag{5.5}$$

$$\underbrace{y_{su} = \sqrt{(1-\rho)}(\sqrt{P_s}h_{su}x + n_{su}),}_{\text{Split signal for information processing}} \tag{5.6}$$

where n_{su} denotes the addictive white Gaussian noise (AWGN) at the UAV, which is modeled as a complex Gaussian random variable with zero mean and variance σ_{su}^2 and x is the transmitted symbol by the source with $\mathbb{E}\{|x|^2\} = 1$, where $\mathbb{E}\{.\}$ and $|.|$ represent the statistical expectation and the norm respectively. Similarly, the signal received at the ith user can be given by

$$y_i = \sqrt{P_u}h_i\hat{x} + n_i, \tag{5.7}$$

where $\hat{x}$ denotes the re-transmit symbol by the UAV, which can be the re-encoded symbol received from the source or the symbol obtained from the cache. Let P_s denote the transmission power of the source node during phase I. Therefore, the amount of energy harvested[3] by the UAV and can be expresses as

$$E_u = \eta\rho T\tau(P_s|h_{su}|^2 + \sigma_{su}^2), \tag{5.8}$$

where η denotes the energy conversion efficiency of the EH circuit. Thus, the transmit power of the UAV during phase II can be given by

$$P_u = \frac{E_u}{\frac{T(1-\tau)}{I}} = \frac{\tau I}{(1-\tau)}\eta\rho(P_s|h_{su}|^2 + \sigma_{su}^2), \tag{5.9}$$

where I denotes the total number of users to be served within the network. The received signal-to-noise ratio (SNR) at the UAV and ith user can be written as

$$\gamma_{su} = \frac{P_s|h_{su}|^2}{\sigma_{su}^2}, \tag{5.10}$$

---

3 Single EH circuit has nonlinear property due to the saturation characteristics of the circuit. Nevertheless, using multiple parallel EH circuits, a large enough linear conversion region can be obtained [23, 24].

and

$$\gamma_i = \frac{P_u |h_i|^2}{\sigma_i^2}. \tag{5.11}$$

For simplicity, it is assumed that $\sigma^2 = \sigma_{su}^2 = \sigma_i^2$. Considering (5.10) and (5.11), and assuming Gaussian codebook, the achievable information rate at the UAV and the ith user can be written as

$$R_u = B \log_2(1 + \gamma_{su}), \tag{5.12}$$

and

$$R_i = B \log_2(1 + \gamma_i), \quad \forall i, \tag{5.13}$$

where B is the channel bandwidth of the communication links.

5.2.3 Caching Mechanism at the UAV

A probabilistic caching method is considered at the UAV similar to the works presented in [16, 20, 25]. Without loss of generality, we assumed that the UAV does not possess any information related to the content popularity. Therefore, the UAV δ portion of the information received from the source, where $(0 \leq \delta \leq 1)$, will be stored in its cache memory. For instance, when user requests information from the main servers, if δ portion of this information is available at the cache, only the remaining portion needs to be sent from the servers, reducing network congestion.

5.3 Optimization Problem Formulation

In this section, two optimization problems are formulated to identify the optimal ρ and τ values, and to generate the optimal trajectory of the UAV to its ending location, which maximizes the serving information rate at the user while assuring predefined QoS constraints.

5.3.1 Maximization of the Achievable Information Rate at the User

The main objective of this section is to maximize the achievable information rate at the users under the predefined QoS constraints considering the caching capacity of the UAV. The corresponding optimization problem (P1) can be written as

(P1) : $\max\limits_{\rho,\tau}$ $\left(\dfrac{T(1-\tau)}{I}\right) B \log_2(1+\gamma_i)$

subject to:

(C1) : $\tau T(R_u + (\delta r_0)) \geq \left(\dfrac{T(1-\tau)}{I}\right) B \log_2(1+\gamma_i),$ (5.14)

(C2) : $P_u^\star T \leq E^\star,$

(C3) : $0 \leq \tau \leq 1,$

(C4) : $0 \leq \rho \leq 1,$

where $P_u \geq (P_u^\star = \psi P_s), 0 < \psi \leq 1$ denotes the QoS threshold of the minimum required transmission power of the UAV toward a single user, $E^\star$ denotes the available energy for communication at the UAV, and r_0 denotes the required information rate by the user. Constraint (C1) asserts that the UAV's buffer is not empty during the operation period. Constraint (C2) is to assure that the UAV cannot surpass the energy usage beyond the given input. It is also noteworthy that the variable $E^\star$ can be modeled in three ways according to the characteristics of the communication networks. First, if the communication capabilities of the UAV powered solely from the harvested energy, then, $E^\star = E_u$, which is the one used in this chapter. Second, if the UAVs transmission power solely depends on the inbuilt battery of the UAV, then, $E^\star = E_{\text{tot}} - E_p$, where E_{tot} is the available energy at the UAV and E_p is the energy required for the propulsion of the UAV. If the UAV's transmission power utilizes the energy from both the EH and the inbuilt battery, then, $E^\star = E_u + (E_{\text{tot}} - E_p)$. This optimization problem is solved via KKT conditions due to its nonlinearity.

Lemma 5.1 *If* $\lambda_1 \neq 0 \Rightarrow \left(\frac{T(1-\tau)}{I}\right) B \log_2(1+\gamma_i) - \tau T(R_u + (\delta r_0)) = 0;$ $\lambda_2 \neq 0 \Rightarrow (P_u^\star - E_u) = 0.$ *Then, we can obtain the following optimal values*

$$\rho = \dfrac{(1-\tau)P_s\psi}{I\eta(P_s * h_{su}^2 + \sigma^2)}, \tag{5.15}$$

$$\tau = \dfrac{ITB \log_2(1+\gamma_i)a_2}{ITB \log_2(1+\gamma_i)a_2 + a_3}, \tag{5.16}$$

where $a_2 = \eta(P_s h_{su}^2 + \sigma^2)$ *and* $a_3 = B \log_2(1+\gamma_u + \delta \ln(2)r_0)$. *It is noteworthy that the obtained closed-form expressions are valid only when the given KKT conditions are satisfied. Otherwise, solution becomes infeasible with respect to the proposed system model. It is also noteworthy that the solution provided in (5.15) and (5.16) for (P1) cannot be guaranteed as the global optimal due to the nonlinearity of the problem. The detailed proof of the optimal solutions are given in Appendix 3.A.*

5.3.2 Trajectory Optimization with Fixed Time and Energy Scheduling

The main objective of this section is to optimize the UAV's trajectory while maximizing the average transmission rate of the users. Achievable average transmission rate at the UAV and the users can be expressed as

$$\bar{R}_u = \frac{1}{N} \sum_{n=1}^{N} \phi_u[n] R_u, \tag{5.17}$$

and

$$\bar{R}_i = \frac{1}{N} \sum_{n=1}^{N} \phi_i[n] R_i, \tag{5.18}$$

where ϕ_u and ϕ_i denote the binary variables, which reflect the UAV's time scheduling during the operation time. If the UAV finishes obtaining information from the source $\phi_u = 1$; otherwise, $\phi_u = 0$. Similarly, if the UAV has finished serving the ith user, $\phi_i = 1$; otherwise, $\phi_i = 0$. Considering the time and power scheduling obtained via (P1), the UAV's trajectory optimization problem can be written as

$$(P2) : \max_{\mathbf{q}} \quad \sum_{i=1}^{I} \bar{R}_i$$

subject to:

$$(C2.1) : \frac{1}{N} \sum_{n=1}^{N} \phi_u[n](\bar{R}_u[n] + \delta[n] r_0[n]) \geq \bar{R}_i,$$

$$(C2.2) : \frac{1}{N} \sum_{n=1}^{N} \phi_i[n] \bar{R}_i[n] \geq r_0[n], \tag{5.19}$$

$$(C2.3) : (x[n] - x_0)^2 + (y[n] - y_0)^2 \leq \frac{V_m T}{N},$$

$$(C2.4) : (x[n+1] - x[n])^2 + (y[n+1] - y[n])^2 \leq \frac{V_m T}{N},$$

$$(C2.5) : (x_d - x[n])^2 + (y_d - y[n])^2 \leq \frac{V_m T}{N},$$

where V_m denotes the maximum velocity of the UAV and x_0, y_0 and x_d, y_d denote the starting and ending locations of the UAV, respectively. Constraint (C2.1) makes sure that the achievable rate at the UAV with the combination of caching scheme is adequate to satisfy user requested information rate. Constraint (C2.2) asserts that the achievable rate at the user always satisfies within the proposed caching scheme. Constraints (C2.3)–(C2.5) represent the mobility constraints of the UAV. The optimization problem (P2) is difficult to solve due to the non-convex constraints (C2.1) and (C2.2) with respect to the UAV's coordination. Thus, using the successive convex optimization technique, the formulated problem approximately converts to a convex optimization problem. First, R_i in (5.13) can be reformulated as

$$\hat{R}_i = B \log_2 \left(1 + \frac{P_u L_0}{\sigma^2 (H + S_x[n] + S_y[n])} \right), \tag{5.20}$$

where $S_{xi} = (x[n] - x_i)^2$, $S_{yi} = (y[n] - y_i)^2$, and H is the flying altitude of the UAV. In order to prove the concavity of the (5.20) with respect to the UAV's trajectory, the following bivariate function [26] is provided:

$$f(\bar{x}, \bar{y}) = \log_2 \left(1 + \frac{J}{D + \bar{x} + \bar{y}} \right), \tag{5.21}$$

where J and D are positive integers, $x \geq 0$ and $y \geq 0$. It can be proved that the Hessian matrix of (5.21) is positive if $\bar{x} \geq 0$ and $\bar{y} \geq 0$. Therefore, the function in (5.21) is convex with respect to $\bar{x}$ and $\bar{y}$. Considering (5.21), $\hat{R}_i$ in (5.20) can also be proved convex with respect to $S_x[n]$ and $S_y[n]$. Thus, optimization problem (P2) can be expressed as a convex optimization problem approximately as follows:

$$(\text{P2.2}) : \max_{q,S} \quad \sum_{i=1}^{I} \bar{R}_i$$

subject to:

$$(\text{C2.2.1}) : \frac{1}{N} \sum_{n=1}^{N} \phi_u[n](\hat{R}_u[n] + \delta[n] r_0[n]) \geq \bar{R}_i,$$

$$(\text{C2.2.2}) : \frac{1}{N} \sum_{n=1}^{N} \phi_i[n] \hat{R}_i[n] \geq r_0[n],$$

$$(\text{C2.2.3}) : (x[n] - x_0)^2 + (y[n] - y_0)^2 \leq \frac{V_{max} T}{N}, \tag{5.22}$$

$$(\text{C2.2.4}) : (x[n+1] - x[n])^2 + (y[n+1] - y[n])^2 \leq \frac{V_{max} T}{N},$$

$$(\text{C2.2.5}) : S_{xi} \leq (x_l[n] - x_i)^2 + 2(x_l[n] - x_i)(x[n] - x_l[n]), \forall n, i,$$

$$(\text{C2.2.6}) : S_{yi} \leq (y_l[n] - y_i)^2 + 2(y_l[n] - y_i)(y[n] - y_l[n]), \forall n, i,$$

$$(\text{C2.2.7}) : S_{xd} \leq (x_l[n] - x_d)^2 + 2(x_l[n] - x_d)(x[n] - x_l[n]), \forall n, i,$$

$$(\text{C2.2.8}) : S_{yd} \leq (y_l[n] - y_d)^2 + 2(y_l[n] - y_d)(y[n] - y_l[n]), \forall n, i.$$

To generate constraints (C2.2.5)–(C2.2.8), the following steps are performed. Considering (5.20), the following inequality can be written:

$$\begin{aligned} S_{xi}[n] &\leq (x[n] - x_i)^2, \\ S_{yi}[n] &\leq (y[n] - y_i)^2. \end{aligned} \tag{5.23}$$

In order to convert the constraints in (5.23) convex, we defined a quadratic function as $f(x) = (x - c)^2$, where c is a constant. Then, by using first-order Taylor expansion, for any given x in iteration number l, we can obtain the following inequality:

$$(x - c)^2 \geq (x_l - c)^2 + 2(x_l - c)(x - x_l). \tag{5.24}$$

Since the quadratic function $f(x) = (x - c)^2$ is convex with respect to x, constraints (C2.2.5)–(C2.2.8) can be easily proved as convex. It is also noteworthy that the optimal objective value of (P2.2) is a lower bound of (P2). Since (P2.2) changed into convex, (5.22) can be solved by using standard convex optimization tools and techniques such as CVX [26]. Thus, an algorithm to optimize the UAV's trajectory with the fixed time and power profiles is given in Algorithm 5.1.

Algorithm 5.1: UAV's Trajectory Optimization

Data: $[x_0, y_0], [x_d, y_d], [x_i, y_i] \forall i, \delta, T, V_{max}$

Result: UAV's optimal trajectory $[\mathbf{x}, \mathbf{y}]$

1 Initialize the UAV's trajectory $[x_n, y_n]$ for $n = 1, \ldots, N$ and set $l = 0$

2 **Repeat**

3 Solve (P1) in (5.14) for given x_l and y_l and obtain ρ^* and τ^*;

4 Solve (P2.2) in (5.22) for given ρ^* and τ^*, and update the trajectory as $x_{n,l+1}$ and $y_{n,l+1}$;

5 Set $l = l + 1$;

6 **Until** Lower bound converges under a predefined threshold;

7 return $[x_{n,l}, y_{n,l}]$.

5.4 Numerical Simulation Results

In this section, performance of the proposed system is evaluated with the aid of numerical simulation results. The system layout of the proposed communication network is illustrated in Figure 5.4 considering an area of 1000 m by 700 m. The source and the destination are located at [0, 300] and [1000, 600], respectively. Four users have been consider in this system setup and they are located in [500, 300], [750, 100], [1000, 300], and [750, 500]. The UAV flies at an altitude of 100 m with the maximum speed of 25 m/s [27]. Unless otherwise stated, the following simulation parameters are considered. Total bandwidth $B = 1$ kHz [28], $\sigma^2 = -119$ dBm, $\alpha = 2$, $L_0 = -60$ dB, $T = 200$ s, $I = 4$, and $N = 200$.

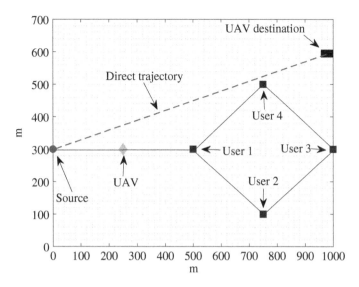

Figure 5.4 System layout for the proposed communication network.

In Figure 5.5, source transmission power versus information rate at the user for various values of caching coefficient δ are provided. It is observed from the figure that the source transmission power and caching coefficient have a significant impact on the information rate at the user. Particularly, when the source transmission power is increased by 500 mW, the information rate at the user increases by approximately 10%. It is also observed that the increase in caching coefficient results in a higher information rate. This observation can be easily explained from the fact that source does not need to send the entire content of the information since the UAV contains partial information within its cache. It is also noteworthy that even with a smaller source transmission power, a similar information rate can be achieved by increasing the caching coefficient. Further, it is noticed from the figure that when there is no caching at the UAV, user requested information rate cannot be achieved for the given range of source transmission power.

A comparison of information rate at the user for different caching coefficients and different T is depicted in Figure 5.6. From the results depicted, it can be observed that there is no effect on the received information rate at the user with the increase in T for a given source transmission power. This is mainly due to the fact that energy and time resource allocation is optimized as in (P1) to maintain the receiving rate at the user such that the achievable throughput can be increased. Furthermore, this phenomena can

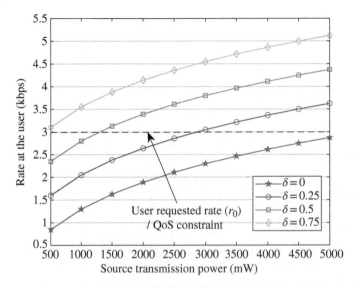

Figure 5.5 Achievable rate at the user versus P_s for various values of caching coefficient δ, where $r_0 = 3$ kbps.

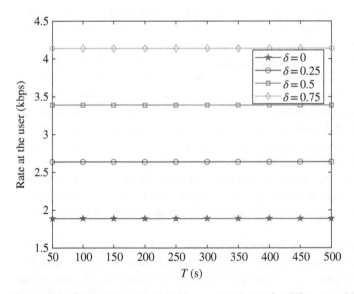

Figure 5.6 Comparison of achievable rate at the user for different caching coefficients and different T.

easily be understood by referring to (5.8), (5.9), (5.15), and (5.16). Similar to Figure 5.5, with increase in cache coefficient, the receiving rate at the users also correspondingly increases.

Next, the impact of transmitted SNR on the information rate at the users for various values of P_s, where $P_s \in \{1500, 1250, 1000\}$ mW, is investigated. The corresponding numerical simulation results are illustrated in Figure 5.7, which are correlated with the (P1) assuming that optimal values for ρ^* and τ^* are chosen as per the solution given in (5.15) and (5.16). It can be clearly seen from the figure that with increase in transmitted SNR, information rate at the users increases exponentially. This observation makes sense since the increase in transmitted SNR results in considerably more harvested energy, leading to an increase in the UAV's transmission power. However, it can also be noticed from the figure that when the transmission power increases with respect to the transmitted SNR, the performance gaps get narrowed down due to the lower noise power.

Figure 5.8 plots the user requested rate r_0 versus optimal values of ρ and τ for different values of $P_s \in \{1000, 1500, 2000\}$ mW. To improve the clarity, each result of similar P_s is given in Figure 5.8a–c. As can be seen clearly from the Figure 5.8, when the requested user information rate increases τ gets increased and ρ gets decreased. This phenomena can be explained referring to Figure 5.3 and (5.12). When the requested rate increases the receiving

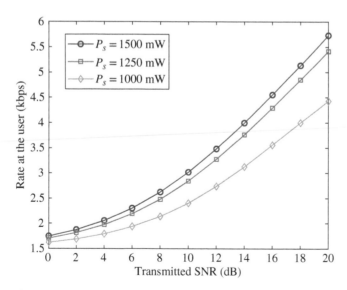

Figure 5.7 Transmission SNR versus achievable rate at the user for different P_s, where $\delta = 0.5$.

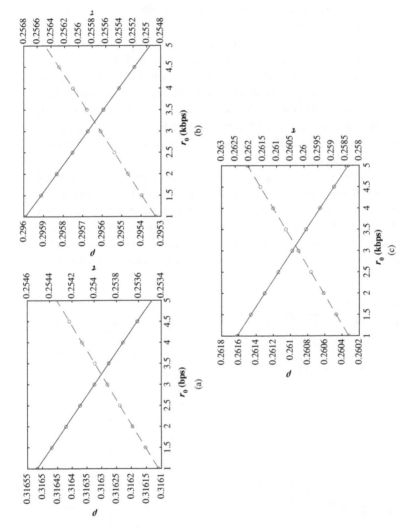

Figure 5.8 User requested rate r_0 versus optimal values of ρ and τ: (a) when the $P_s = 1000\,\text{mW}$, (b) when the $P_s = 1500\,\text{mW}$, and (c) when $P_s = 2000\,\text{mW}$.

SNR at the user needs to be increased correspondingly. Thus, more energy at the UAV needs to be harvested while having an equal better portion of the signal for information processing. Then, both ρ and τ jointly need to be selected to satisfy the aforementioned requirement. It can also be noticed from the Figure 5.8 that the data points of ρ get higher compared to the sub-figures when the transmission power decreases.

The Bird's-eye view of the optimal trajectory of the UAV obtained via Algorithm 5.1 is presented in Figure 5.9 with the direct and sub-optimal trajectories. Sub-optimal trajectory is designed to follow to direct trajectories between the communication nodes. The serving order of users for the given scenario is user1, user2, user3, and user4. It can be observed from Figure 5.9 that the UAV prefers to fly following a trajectory closer to the sub-optimal trajectory to achieve maximum rate at the users. It can also be seen that the UAV's travelling distance is lower than the sub-optimal trajectory, which can lead to improved energy efficiency than the sub-optimal solution. Even though direct trajectory is more energy efficient, only user4 can get the higher rate while the rest of the users experience lower information rates. Information rate at the users for each trajectory is presented in Table 5.2 for better understanding with relation to different trajectories. The values presented in Table 5.2 prove that the optimal trajectory obtained

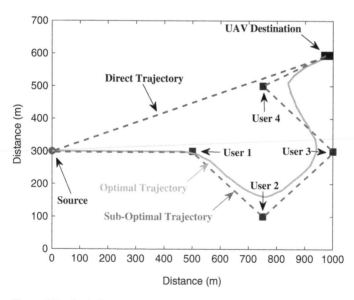

Figure 5.9 Optimized trajectory of the UAV for the given communication setup with four users, where $P_s = 2000\,\text{mW}$ and $\delta = 0.5$.

Table 5.2 Rate at the users for different UAV's trajectories.

Rate (kbps)	user1	user2	user3	user4
Direct	3.1466	0.7664	1.8251	4.4251
Sub-optimal	4.2791	4.2851	4.2938	4.2878
Optimal (Algo 5.1)	4.4251	4.4765	4.4251	4.2959

via Algorithm 5.1 shows superior performance compared to the direct and sub-optimal trajectories.

5.5 Conclusion

In this chapter, a self-energized UAV-assisted caching relaying scheme is investigated. The UAV's communication capabilities were powered solely by the PS-SWIPT EH technique and it employs DF relaying protocol to assist information transmission to the users from the source node. A new transmission block diagram has been developed to accommodate communication processes within the system. Then, the problem of identifying optimal time and energy resources for the communication system and the optimal UAV's trajectory are addressed while adhering to the QoS requirements of the communication network. The corresponding results of the theoretical analysis are provided with the aid of numerical simulation results to identify the impacts of the system parameters on its information rate at the users. This work can be further extended to many captivating research directions by including modifications such as use of multiple UAVs (i.e. UAV swarms) without channel state information at communication nodes, nonlinear EH model, and integrating full-duplex-enabled UAVs.

Acknowledgments

The authors would like to thank the Competitiveness Enhancement Program of the National Research Tomsk Polytechnic University, Russia, Grant No. VIU-ISHITR-180/2020, and Russian Foundation for Basic Research (RFBR), project No. 19-37-90037, for supporting this work.

Appendix 5.A
Proof of Optimal Solutions Obtained in (P1)

The corresponding Lagrangian expression to (P1) can be written as

$$\mathcal{L}(\rho, \tau; \lambda_1, \lambda_2, \lambda_3, \lambda_4) = M(\rho, \tau) - \lambda_1 N(\rho, \tau) - \lambda_2 O(\rho, \tau)$$
$$- \lambda_3 P(\rho, \tau) - \lambda_4 Q(\rho, \tau), \tag{5.A.1}$$

where $M = \left(\frac{T(1-\tau)}{I}\right) B \log_2(1 + \gamma_i)$, $N = M - \tau T(R_u + (\delta r_0)) \le 0$, $O = (T P_u^\star - E_u) \le 0$, $P = (\rho - 1) \le 0$ and $Q = (\tau - 1) \le 0$. To satisfy the local optimality condition $\nabla \mathcal{L}(\rho, \tau; \lambda_1, \lambda_2, \lambda_3, \lambda_4) = 0$, the equations can be expressed as

$$\frac{\partial \mathcal{L}(\rho, \tau; \lambda_1, \lambda_2, \lambda_3, \lambda_4)}{\partial \rho} = \frac{\partial M}{\partial \rho} - \frac{\partial \lambda_1 N}{\partial \rho} - \frac{\partial \lambda_2 O}{\partial \rho} - \frac{\partial \lambda_3 P}{\partial \rho} - \frac{\partial \lambda_4 Q}{\partial \rho} = 0,$$
$$\tag{5.A.2}$$

$$\frac{\partial \mathcal{L}(\rho, \tau; \lambda_1, \lambda_2, \lambda_3, \lambda_4)}{\partial \tau} = \frac{\partial M}{\partial \tau} - \frac{\partial \lambda_1 N}{\partial \tau} - \frac{\partial \lambda_2 O}{\partial \tau} - \frac{\partial \lambda_3 P}{\partial \tau} - \frac{\partial \lambda_4 Q}{\partial \tau} = 0.$$
$$\tag{5.A.3}$$

The complementary slackness expression of feasibility conditions given in (5.A.1) can be written as $\lambda_1 N(\rho, \tau) = \lambda_2 O(\rho, \tau) = \lambda_3 P(\rho, \tau) = \lambda_4 Q(\rho, \tau) = 0$. The conditions for non-negativity can be represented as $\rho, \tau, \lambda_1, \lambda_2, \lambda_3, \lambda_4 \ge 0$. Following is the identified feasible KKT conditions for the corresponding Lagrangian. All other KKT conditions are not feasible due to its contradictory behavior compared to the proposed energy efficient caching scheme.

Conditions: $\lambda_1 \ne 0 \Rightarrow \left(\frac{T(1-\tau)}{I}\right) B \log_2(1 + \gamma_i) - \tau T(R_u + (\delta r_0)) = 0$; $\lambda_2 \ne 0$ $\Rightarrow (P_u^\star - E_u) = 0$. Therefore, by using (5.A.2) following equations can be obtained.

$$-\frac{P_s \eta \tau T h_{su}^2 h_i^2}{\sigma^2 \log_2(1 + \gamma_i)} - \lambda_1 \frac{P_s \eta \tau T h_{su}^2 h_i^2}{\sigma^2 \log_2(1 + \gamma_i)} - \lambda_2 \frac{I P_s \eta \tau h_{su}^2}{(1 - \tau)} = 0, \tag{5.A.4}$$

$$-\left(\frac{ITB \log_2(1 + \gamma_i)}{\log(2)} - \frac{I^2 BTP_s \eta \rho h_{su}^2 h_i^2 \left(1 + \frac{\tau}{(1-\tau)}\right)}{\sigma^2 \log_2(1 + \gamma_i)}\right) - \lambda_1 \left(\frac{ITB \log_2(1 + \gamma_i)}{\log(2)}\right.$$

$$-\frac{I^2 P_s BT \eta \rho h_{su}^2 h_i^2 \left(1 + \frac{\tau}{(1-\tau)}\right)}{\sigma^2 \log_2(1 + \gamma_i)} - \frac{TB \log_2(1 + \gamma_u)}{\log(2)} + T\delta r_0$$

$$- \lambda_2 \left(\frac{I P_s \eta \rho h_{su}^2}{(1 - \tau)} \left(1 + \frac{\tau}{1 - \tau}\right)\right) = 0, \tag{5.A.5}$$

$$P_s \eta \tau \rho h_{su}^2 \left(\frac{I}{(1-\tau)-1} \right) = 0, \tag{5.A.6}$$

$$\frac{T(1-\tau)}{I} B \log_2(1+\gamma_i) - T\tau(B \log_2(1+\gamma_{su}) + \delta r_0) = 0. \tag{5.A.7}$$

Considering (5.A.4) and (5.A.5), λ_1 and λ_2 can be expressed as

$$\lambda_1 = \frac{(\tau-1)\ln(2)^2 h_i^2 (-x_1 - x_2)}{\sigma^2 B \log_2(1+\gamma_i)(x_2(1+\ln(2)) + x_1 + x_3)}, \tag{5.A.8}$$

$$\lambda_2 = \frac{x_1 + x_2}{x_2(1+\ln(2)) + x_1 x_3}, \tag{5.A.9}$$

where $x_1 = \sigma^2(\tau-1)B \log_2(1+\gamma_i)^2$, $x_2 = I\rho\eta P_s \ln(2)^3 h_{su}^2 h_i^2$ and $x_3 = (1+I)B \log_2(1+\gamma_i) + I\ln(2)^2 \delta r_0$. Then, using (5.20) (5.21) in (5.16) (5.17) considering (5.A.6) (5.19), the optimal solutions for ρ and τ can be given as

$$\rho^* = \frac{(1-\tau)P_s \psi}{I\eta(P_s * h_{su}^2 + \sigma^2)}; \quad \tau^* = \frac{ITB \log_2(1+\gamma_i)a_2}{ITB \log_2(1+\gamma_i)a_2 + a_3}, \tag{5.A.10}$$

where $a_2 = \eta(P_s h_{su}^2 + \sigma^2)$ and $a_3 = B \log_2(1+\gamma_u + \delta \ln(2)r_0)$.

Bibliography

1 Brown, A. and Anderson, D. (2020). Trajectory optimization for high-altitude long-endurance UAV maritime radar surveillance. *IEEE Transactions on Aerospace and Electronic Systems* 56 (3): 2406–2421.

2 Zhang, S., Zhang, H., He, Q. et al. (2018). Joint trajectory and power optimization for UAV relay networks. *IEEE Communications Letters* 22 (1): 161–164.

3 Zeng, Y. and Zhang, R. (2017). Energy-efficient UAV communication with trajectory optimization. *IEEE Transactions on Wireless Communications* 16 (6): 3747–3760.

4 Jayakody, D.N.K., Perera, T.D.P., Ghrayeb, A., and Hasna, M.O. (2020). Self-energized uav-assisted scheme for cooperative wireless relay networks. *IEEE Transactions on Vehicular Technology* 69 (1): 578–592.

5 Wu, J., Li, L., and Du, L. (2020). UAV-assisted relaying transmission design and optimization for high-speed moving sources. *IEEE Access* 8: 195857–195869.

6 Kim, T. and Qiao, D. (2020). Energy-efficient data collection for IoT networks via cooperative multi-hop UAV networks. *IEEE Transactions on Vehicular Technology* 69 (11): 13796–13811.

7 Liu, X., Li, Z., Zhao, N. et al. (2019). Transceiver design and multihop D2D for UAV IoT coverage in disasters. *IEEE Internet of Things Journal* 6 (2): 1803–1815.

8 Zhang, H., Sung, D.K., and Wang, J. (2020). Modelling and analysis of coverage for unmanned aerial vehicle base stations. *IET Communications* 14 (17): 2878–2888.

9 Saxena, V., Jaldén, J., and Klessig, H. (2019). Optimal UAV base station trajectories using flow-level models for reinforcement learning. *IEEE Transactions on Cognitive Communications and Networking* 5 (4): 1101–1112.

10 Perera, T.D.P., Jayakody, D.N.K., Sharma, S.K. et al. (2018). Simultaneous wireless information and power transfer (SWIPT): recent advances and future challenges. *IEEE Communications Surveys Tutorials* 20 (1): 264–302.

11 Jayakody, D.N.K., Thompson, J., Chatzinotas, S., and Durrani, S. (2017). *Wireless Information and Power Transfer: A New Paradigm for Green Communications*. Springer.

12 Zhou, S., Gong, J., Zhou, Z. et al. (2015). GreenDelivery: proactive content caching and push with energy-harvesting-based small cells. *IEEE Communications Magazine* 53 (4): 142–149.

13 Kumar, A. and Saad, W. (2015). On the tradeoff between energy harvesting and caching in wireless networks. *2015 IEEE International Conference on Communication Workshop (ICCW)*, pp. 1976–1981.

14 Niyato, D., Kim, D.I., Wang, P., and Song, L. (2016). A novel caching mechanism for internet of things (IoT) sensing service with energy harvesting. *2016 IEEE International Conference on Communications (ICC)*, pp. 1–6.

15 Zhang, S., Zhang, N., Fang, X. et al. (2017). Cost-effective vehicular network planning with cache-enabled green roadside units. *2017 IEEE International Conference on Communications (ICC)*, pp. 1–6.

16 Gautam, S., Vu, T.X., Chatzinotas, S., and Ottersten, B. (2019). Cache-aided simultaneous wireless information and power transfer (SWIPT) with relay selection. *IEEE Journal on Selected Areas in Communications* 37 (1): 187–201.

17 Chen, M., Saad, W., and Yin, C. (2019). Echo-liquid state deep learning for 360° content transmission and caching in wireless VR networks with cellular-connected UAVs. *IEEE Transactions on Communications* 67 (9): 6386–6400.

18 Zhao, N., Cheng, F., Yu, F.R. et al. (2018). Caching UAV assisted secure transmission in hyper-dense networks based on interference alignment. *IEEE Transactions on Communications* 66 (5): 2281–2294.

19 Chai, S. and Lau, V.K.N. (2020). Online trajectory and radio resource optimization of cache-enabled UAV wireless networks with content and energy recharging. *IEEE Transactions on Signal Processing* 68: 1286–1299.

20 Zhou, F., Wang, N., Luo, G. et al. (2020). Edge caching in multi-UAV-enabled radio access networks: 3D modeling and spectral efficiency optimization. *IEEE Transactions on Signal and Information Processing over Networks* 6: 329–341.

21 Iskandar, S.S. and Shimamoto, S. (2005). The channel characterization and performance evaluation of mobile communication employing stratospheric platform. *IEEE/ACES International Conference on Wireless Communications and Applied Computational Electromagnetics, 2005,* pp. 828–831.

22 You, C. and Zhang, R. (2019). 3D trajectory optimization in Rician fading for UAV-enabled data harvesting. *IEEE Transactions on Wireless Communications* 18 (6): 3192–3207.

23 Ma, G., Xu, J., Zeng, Y., and Moghadam, M.R.V. (2019). A generic receiver architecture for MIMO wireless power transfer with nonlinear energy harvesting. *IEEE Signal Processing Letters* 26 (2): 312–316.

24 Hu, Z., Zheng, C., He, M., and Wang, H. (2020). Joint Tx power allocation and Rx power splitting for SWIPT systems with battery status information. *IEEE Wireless Communications Letters* 9 (9): 1442–1446.

25 Vu, T.X., Chatzinotas, S., and Ottersten, B. (2018). Edge-caching wireless networks: performance analysis and optimization. *IEEE Transactions on Wireless Communications* 17 (4): 2827–2839.

26 Boyd, S., Boyd, S.P., and Vandenberghe, L. (2004). *Convex Optimization.* Cambridge University Press.

27 Meredith, J. (2017). Study on Enhanced LTE Support for Aerial Vehicles. *Tech. Rep. No. 36.777* 3GPP.

28 Yang, G., Lin, X., Li, Y. et al. (2018). A telecom perspective on the internet of drones: from LTE-advanced to 5G. *arXiv preprint arXiv:1803.11048.*

6

Performance of mmWave UAV-Assisted 5G Hybrid Heterogeneous Networks

Muhammad K. Shehzad, Muhammad W. Akhtar, and Syed A. Hassan

School of Electrical Engineering and Computer Science (SEECS), National University of Sciences and Technology (NUST), Islamabad, Pakistan

6.1 The Significance of UAV Deployment

Unmanned aerial vehicles (UAVs) can be used as airborne base stations because of their high mobility, autonomous deployment, and cost-effective characteristics. Also, by integrating UAVs with the existing cellular networks, they can provide improved coverage [1–4]. The application of UAV in millimeter wave (mmWave) frequencies, in particular, can provide future wireless networks with low latency, high data rate, and seamless network coverage [5]. On the other hand, it is well known that mmWave signals are vulnerable to blockage because of their low diffraction.

As a potential technology to enable emerging ad hoc communications through the operation of UAVs for terrestrial and low-altitude users, UAV-assisted communication networking has acquired considerable research interest [6–8]. In various applications, UAVs can be used as base stations (BSs) to boost signal-to-noise ratio (SNR), wider scope, better energy consumption, and better protection. Therefore, apart from conventional terrestrial BSs, because of their cost-effectiveness and the ability to dynamically change their altitude and locations, UAV BSs are more beneficial.

In particular, the use of UAV swarm as a network can enable users to improve SNR by obtaining a line-of-sight (LoS) path compared to land-based BSs, as terrestrial BSs often face physical obstacles, including buildings, trees, and hills [9]. Their potential to have fast and infrastructure-less rollout in next-generation networks to meet high-speed data requirements gives UAVs essential significance. Current UAV research has focused on several directions, such as its implementation for the supply of goods, the creation

Autonomous Airborne Wireless Networks, First Edition.
Edited by Muhammad Ali Imran, Oluwakayode Onireti, Shuja Ansari, and Qammer H. Abbasi.
© 2021 John Wiley & Sons Ltd. Published 2021 by John Wiley & Sons Ltd.

of an ad hoc and rapid network for disastrous infrastructure, supply chain, and traffic monitoring and management. The ability to adjust elevation and position is a distinguishing aspect of a UAV-assisted communication network.

UAVs are expected to support different fifth-generation (5G) wireless network implementations. They can act as local hot spots in densely populated regions, help improve wireless users' service reliability by acting as a content cache, and meet first responders' connectivity requirements at any time/location when they are urgently needed [10, 11]. UAVs can switch easily from one place to another due to their flexible, three-dimensional (3D) mobility in free space to provide on-demand communication assistance. They can change their locations to avoid blockage/shadowing and can help incorporate data relay/transport implementations.

6.2 Contribution

This work highlights the role of mmWave and teraHertz (THz) technology in the domain of UAV communication. Additionally, we shed light on some of the critical issues related to UAV communication along with potential applications. In particular, the role of UAVs as aerial hubs to provide fronthaul connectivity between the small-cell base stations (SCBs) and core-network is investigated in detail. Specifically, the association of SCBs is presented by considering multiple communication-related constraints, for instance, SNR, bandwidth, data rate, and backhaul limitations. Also, we provide an association algorithm, which is heuristic in nature. Moreover, to consider backhaul data rate limitation, an algorithm is addressed, which not only considers the backhaul rate limitation but also provides fairness to the overall network.

6.3 The Potential of mmWave and THz Communication

The use of mmWave bands may be very exciting for UAV communications to support data rate requirements for mobile applications that require high throughput. In particular, UAVs can retain LoS or at least a reasonable non-line-of-sight (NLoS) connection with the desired user by hovering at a favorable spot, which is critical to ensure reliable connection on mmWave frequencies due to high path losses [12, 13]. The 28 GHz licensed bands and the 60 GHz unlicensed bands are recommended for future high-speed UAV communications in mmWave [14].

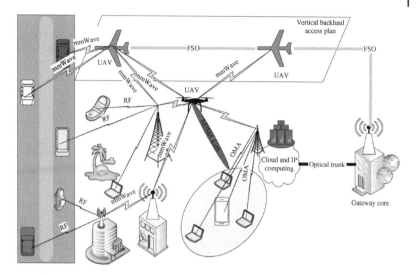

Figure 6.1 An illustration of mmWave/THz and UAVs integrated hybrid communication network.

As in Figure 6.1, UAVs acting as flying BSs or remote relay backhaul nodes may be configured with cellular networks to increase coverage in a heterogeneous wireless network. UAVs can also be used to assist mmWave communications as prime targets. Free space optics (FSO) can be used as a link between UAVs and gateway exchange. The installation of beam-steerable antenna arrays on a compact UAV with a minimum payload is an efficient way of maintaining connectivity with large capacities considering its small wavelength. The latest access techniques such as non-orthogonal multiple access (NOMA) can also be implemented in UAV-assisted communication systems due to their ability to achieve high SNR. Indeed, the receiver should be positioned inside the transmitter's LoS to completely leverage the benefit of mmWave frequencies [14].

UAVs provided with mmWave facilities can provide on-the-fly communications and build stable LoS connectivity between each other and the ground BSs because of the implausibility of maintaining long LoS links, particularly in densely populated areas. The use of THz band to enable connectivity is seen as a new paradigm with tremendous potential to have enormous choices for bandwidth. The THz band tackles the spectrum scarcity issue and can be used for different applications [15].

The current backhaul technologies consisting of point-to-point (P2P) microwave connectivity and optical fiber are not viable in terms of their related costs as the number of small cells operating at high-frequency

Table 6.1 Impact on the characteristics of signals at THz and mmWave frequency bands [16, 17].

Technologies	THz	mmWaves
Frequency range	100 to 10 GHz	30–300 GHz
Atmospheric losses	High	Medium
Directivity	High	High
Power consumption	Medium requirement	Medium requirement
Network characteristics	P2P	P2MP

band rises. However, in terms of spontaneous movements of a UAV due to hovering at variable altitudes and also mmWave/THz propagation characteristics, the communication medium for airborne mmWave links should be differentiated to take full advantage of mmWave-enabled UAV communication characteristics.

In addition, the communication pattern between transmitter and receiver suffers substantial misalignment due to directionality requirements of transmission in the mmWave frequency band. In particular, spontaneous UAV movements may result in antenna gain misalignment between the receiver and transmitter, which can cause SNR variations at the receiving end, which can also dramatically reduce the system's reliability. To prevent the potential loss in the transmission at the mmWave, the radiation pattern of the antenna in the transceivers should be optimally configured to provide a reliable mmWave connection for UAVs at various levels of their location.

Similarly, it is important to provide accurate channel modeling for UAV-based communication networks operating at mmWave frequencies, which can incorporate the path loss, fading, and variation in the directionality of the antenna in the UAV along with the variation in the altitude of the UAV [9]. Table 6.1 shows the characteristics of mmWave and THz spectrum on the signal propagation, typically the signal characteristics in these bands.

6.4 Challenges and Applications

In this section, we give an overview of the technical challenges for implementing mmWave/THz for UAV communications. Later on, we present some applications of UAV communications.

6.4.1 Challenges

mmWave and THz band communication are the potential solutions to the spectrum scarcity in the conventional microwave communication systems [14]. This section gives an overview of challenges in the deployment of high data rate, low latency, and infrastructure-less UAVs communication in mmWave and THz frequency bands.

6.4.1.1 Complex Hardware Design

Because of the wideband frequency availability and constantly changing channel conditions, a robust hardware design is required, which can track the continuously changing channel in formations and at the same time predict the channel in the next location [18]. A high-frequency band requires high-speed analog-to-digital (A/D) and digital-to-analog (D/A) converters, which may lead to high cost.

6.4.1.2 Imperfection in Channel State Information

UAVs commonly undergo unwanted transitions and rotations, which can often be classified as high-frequency jitter and low-frequency fluttering, because of the atmospheric noise [19]. UAV jitter typically has a high-frequency function and is thus irregular and dangerous for communication. The movement of a UAV is usually slow, irregular, and random. Because of the directivity of mmWave beams, these features have an important effect on the efficiency of mmWave-UAV communications. Channel modeling and the development of stable connections in UAV-to-UAV and/or UAV-to-ground communications infrastructure face new challenges.

6.4.1.3 High Mobility

UAVs typically perform high-speed navigation, unlike terrestrial networks. This allows the channel to shift quickly in mmWave-UAV communications and can cause misalignment of the beam due to strong UAV mobility and repeated handoff of BS [20]. In comparison, a UAV's rapid movement results in extreme Doppler frequency shift and scatter, which is more dominant for high-frequency bands. There is also a long way to go to standardize and implement UAV communications in the mmWave frequency range. Further investigation is needed for the issues that arose from the UAV flight, such as user location search, fast channel estimation, successful beam training and monitoring, and Doppler effect [21].

6.4.1.4 Beam Misalignment

Versatile flexibility makes it ideal for a UAV to optimize signal strength and maximize communication link quality. However, propagation and beamforming are closely coupled in mmWave-UAV communications. For different UAV locations, the channels have unique attributes in both the amplitude and angular domain, which makes the beamforming architecture and multiple access highly critical. In addition, mmWave-UAV communication should be taken into consideration, unlike lower frequency UAV communication using omnidirectional antennas. A UAV aims to obtain knowledge of the channel and environment to find the optimum location. In comparison, the adaptive deployed mmWave-UAV implementation is more complex, where the trajectory and beamforming in real-time can be jointly planned.

In Section 6.4.2, we shed light on the potential applications of UAV in mmWave and THz frequency bands.

6.4.2 Applications

UAV communication, by any measure, is of pivotal importance in most applications of daily life. It has paved the way for the modern system of cellular communication, in particular. In the modern era, there are several uses of UAVs, such as in public safety networks, military operations, agriculture, construction industry, smart transportation systems, and communication with people in regions affected by disasters in the event of a tsunami, earthquake, etc. [2, 22–24].

Also, UAVs can be used as aerial base stations (ABSs) and aerial relays in cellular networks for coverage and capacity enhancement, multiple-input multiple-output (MIMO) and mmWave communication, information dissemination in terrestrial networks, Internet-of-things (IoT) networks for data collections or communication, cache-enabled UAVs, cellular-connected drones as user equipments (UEs), flying ad hoc networks, and smart cities [23, 25]. More importantly, UAVs usage in fronthaul networks has also gained a lot of interest [8, 24, 26, 27], which is the major focus of this chapter. Therefore, in the following, we will discuss the usage of UAVs for fronthaul connectivity and some of the issues associated with such a network.

Among others, one of the promising applications of UAVs is to provide fronthaul connectivity between the terrestrial SCBs and the ground core-network (GCN) [26, 27]. Traditionally, the traffic between the GCN and the SCBs is routed via optical communication. Nevertheless, despite the benefit of obtaining high bandwidth using a fiber-based communication link, the downside of such a technology is high capital expenditure cost (CAPEX) and the deployment time. To overcome these issues, Alzenad et al. [27] introduced a scalable idea of replacing the terrestrial backhaul network with an aerial network. However, such a network brings challenges such as UAV placement, air-to-ground (ATG) channel modeling, an association of SCBs with the UAVs, and hover time optimization of UAVs. Among these, this work focuses on the two challenges, i.e. placement of UAVs and the association of SCBs.

In the following, we present the communication environment for using the UAVs as aerial hubs to provide fronthaul connectivity to SCBs. Later on, we address the placement of UAVs and the association of SCBs.

6.5 Fronthaul Connectivity using UAVs

Consider a UAV-assisted communication environment, as depicted in Figure 6.2. Mainly, there are three network entities, that is, UAVs, SCBs, and a GCN. UAVs are categorized into two classes: child-UAVs and a parent-UAV. Child-UAVs are hovering at an altitude, denoted by h, from the ground level. Also, the parent-UAV, which acts as a communication hub between the child-UAVs and the GCN, is placed at a higher altitude than the child-UAVs so that perfect LoS communication between the GCN and the child-UAVs could become possible. The communication between UAVs and GCN is based on FSO,[1] which aims to provide a high-speed wireless link [28, 29]. Besides, SCBs aggregate and route the uplink/downlink traffic of cellular users via these UAVs to the GCN. Additionally, child-UAVs are capable of sharing the control information, for instance, signal-to-interference-plus-noise ratio (SINR), bandwidth

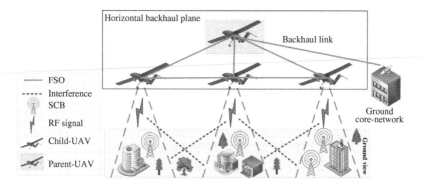

Figure 6.2 Pictorial representation of UAV-enabled wireless network.

---

1 In our model, we assume this link ideal, which means that no losses are taken into account.

and demanded data rate of SCBs, etc., with each other and the parent-UAV. Nonetheless, every child-UAV is responsible for sharing its aggregated information with the parent-UAV, where the parent-UAV considers the limitations of the backhaul link. In addition, the communication between the SCBs and the child-UAVs is radio frequency (RF) based, for which we will describe the communication model later in this chapter.

For the sake of notational convenience, let us denote the coordinates of SCBs and the child-UAVs as (x_m, y_m) and (x_n, y_n, h_n), respectively, where $m = \{1, 2, 3, \dots, M\}$ and $n = \{1, 2, 3, \dots, N\}$. Before moving toward presenting the communication between the SCBs and the UAVs, below, we first deliver the distribution of SCBs, and then we address the placement of UAVs.

6.5.1 Distribution of SCBs

Considering the practical environment, the distribution of SCBs is obtained using a *Matern type-I* hard-core process [30] with the average density of ρ/m^2; therefore, the resulting process gives the average number of SCBs as

$$M = \rho \cdot \exp\left(-\rho\pi d_{\min}^2\right) \cdot S, \tag{6.1}$$

where S denotes the area in which the SCBs are deployed. In addition, $d_{\min}$ represents the minimum distance between the two SCBs. Therefore, it can be said that the average number of SCBs in a region is the product of average density, ρ, and the total area.

6.5.2 Placement of UAVs

In the literature, there exist many papers, e.g. [8, 26, 31, 32, 34–36], to solve the placement of UAVs. However, the placement of UAVs considering the distribution of cellular users, which aims to maximize the sum rate or number of served users, etc., is an optimization task. Notably, such an optimization problem can be non-convex, non-smooth, and nondeterministic polynomial (NP)-hard [37]. Therefore, in this work, for the sake of simplicity, we take advantage of using the k-means clustering algorithm [38] for the positioning of UAVs. Nonetheless, interested readers are encouraged to read, e.g. [8, 26, 31, 33].

The k-means clustering algorithm partitions the M SCBs into k regions ($k < M$), represented by $\mathcal{R} = \{R_1, R_2, \dots, R_k\}$. Within k regions, an SB $m \in M$ belongs to a region having the nearest mean. Mathematically,

$$\underset{\mathcal{R}}{\mathrm{argmin}} \sum_{j=1}^{k} \sum_{m \in R_j} \|m - \mathbf{C}_j\|^2, \tag{6.2}$$

where $\mathbf{C}_j$ represents the centroid (mean) of all the points in R_j, which is the two-dimensional (2D) position of a child-UAV in that region. Therefore, N UAVs are positioned on the obtained centroids, with fixed height for all UAVs. Then, the communication takes place between the SCBs and the UAVs, which we discuss in the following section.

6.6 Communication Model

Consider the stochastic geometry, depicted in Figure 6.3; the horizontal distance between the mth SCB and the nth UAV is represented as

$$d_{m,n} = \sqrt{(x_m - x_n)^2 + (y_m - y_n)^2}. \tag{6.3}$$

In addition, the probability of LoS between the mth SCB and the nth UAV is a fundamental factor in the path loss calculation. Following Al-Hourani et al. [39], the LoS probability is given by

$$\zeta_{m,n}^{\ell} = \frac{1}{1 + a \cdot \exp\{-b(\phi - a)\}}, \tag{6.4}$$

where a and b are environment constants, and ϕ is the angle (in degrees) between a UAV and an SCB. Further, the probability of NLoS is written as

$$\zeta_{m,n} = 1 - \zeta_{m,n}^{\ell}. \tag{6.5}$$

Furthermore, the angle of elevation, ϕ, between an SCB and a UAV is calculated as

$$\phi = \arctan\left(\frac{h_n}{d_{m,n}}\right). \tag{6.6}$$

The ATG path loss model, given in [39], along with fading, δ, is presented as

$$\mathbb{G}_{m,n} = \Lambda_0 + \zeta_{m,n}^{\ell} \cdot \varepsilon^{\ell} + \zeta_{m,n} \cdot \varepsilon - \delta, \tag{6.7}$$

where

$$\Lambda_0 = 10 \log_{10}\left(\frac{4\pi \cdot \upsilon_{m,n}}{\lambda_c}\right)^{\gamma}, \tag{6.8}$$

which is the free space path loss (FSPL), and γ denotes the path-loss exponent. Also, ε^{ℓ} and ε are the attenuation factors for the LoS and NLoS communication links, respectively. Moreover,

$$\upsilon_{m,n} = \sqrt{h_n^2 + d_{m,n}^2}, \tag{6.9}$$

which represents the distance between a UAV and an SCB. Moreover, δ is expressed as

$$\delta[\mathrm{dB}] = \zeta_{m,n}^{\ell} \cdot \xi_0 + \zeta_{m,n} \cdot \xi_1, \tag{6.10}$$

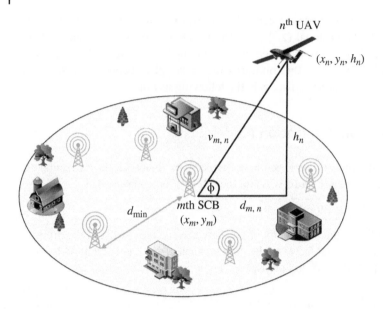

Figure 6.3 Stochastic geometry for the communication between an SCB and a UAV.

where the envelopes of ξ_0 and ξ_1 follow *Nakagami* distributions. In other words, $|\xi_\iota| \sim Nakagami(\iota)$, where $\iota = \{0, 1\}$, and ι is the shape parameter, which takes the value 1 for Rayleigh fading and 4 for Rician fading.

By considering the path loss model given above, the received power, $P_{m,n}^r$, at the mth SCB is written as

$$P_{m,n}^r = P_n^t - \mathbb{G}_{m,n},\tag{6.11}$$

where P_n^t represents the transmitted power of the nth UAV. Finally, assuming the omnidirectional antenna at the UAV, the SINR (Γ) at the mth SCB is determined as

$$\Gamma_{m,n} = \frac{P_{m,n}^r}{\sigma_{\text{noise}}^2 + I},\tag{6.12}$$

where σ_{noise}^2 is the noise power. Also, I represents the sum of the interference from the remaining $(N - 1)$ UAVs, as we assumed omnidirectional antennas at the UAVs. In addition, the required bandwidth of the mth SCB from the nth UAV is calculated as

$$w_{m,n} = \frac{r_{m,n}}{\log_2(1 + \Gamma_{m,n})},\tag{6.13}$$

where $r_{m,n}$ is the requested data rate of mth SCB from nth UAV.

6.6.1 Communication Constraints and Objective

Considering the distribution of SCBs and fixed location of child-UAVs (discussed in Section 6.5.2), communication links are optimized. Nevertheless, the communication between the SCBs and UAVs is limited by a number of factors, which are discussed below.

1. To avoid overloading, a child-UAV, n, can accommodate a maximum of $L^{\max}$ links.
2. There is a maximum bandwidth limit, W, which a child-UAV can distribute among its candidates (SCBs).
3. To maintain quality-of-service (QoS) requirements, the minimum SINR, $\Gamma^{\min}$, criteria is taken into account, as it is of prime importance for the distribution of bandwidth.
4. The mth SCB will only be served by a particular nth child-UAV.
5. Besides, to maintain the quality of backhaul link, which is the communication path between the parent-UAV and the GCN, a maximum backhaul data rate limit, R_B, is considered.

Keeping in view the above communication constraints, the objective of this work is to find the best possible association of SCBs with UAVs such that the sum rate of the overall system can be maximized. The association of SCBs with UAVs is dependent on communication factors, i.e. $L^{\max}$, W, $\Gamma^{\min}$, and R_B. Let $\mathbf{A}$, with dimension $M \times N$, be the association matrix, where the rows and columns of matrix $\mathbf{A}$ represent the SCBs and UAVs, respectively. In addition, let $a_{m,n} \in \{0,1\}$ be the (m,n)th entry of matrix $\mathbf{A}$. Hence, the objective, i.e. association of SCBs with UAVs is, mathematically formulated as

$$\max_{\{a_{m,n}\}} \sum_{m=1}^{M} \sum_{n=1}^{N} r_{m,n} \cdot a_{m,n} \tag{6.14}$$

s.t.

$$\sum_{m=1}^{M} w_{m,n} \cdot a_{m,n} \leq W_n, \qquad \forall n \tag{6.15}$$

$$\sum_{m=1}^{M} a_{m,n} \leq L_n^{\max}, \qquad \forall n \tag{6.16}$$

$$\Gamma_{m,n} \cdot a_{m,n} \geq \Gamma^{\min}, \qquad \forall m,n \tag{6.17}$$

$$\sum_{n=1}^{N} a_{m,n} \leq 1, \qquad \forall m \tag{6.18}$$

$$\sum_{m=1}^{M} \sum_{n=1}^{N} r_{m,n} \cdot a_{m,n} \leq R_B, \tag{6.19}$$

where $a_{m,n}$ is the optimization parameter. From (6.14),

$$S_r = \sum_{m=1}^{M} \sum_{n=1}^{N} r_{m,n} \cdot a_{m,n}, \tag{6.20}$$

where S_r represents the total achieved sum rate from the overall network.

In constraint (6.15), W_n is the maximum available bandwidth to nth child-UAV, and $w_{m,n}$ is the required bandwidth of the mth SCB from nth child-UAV. Additionally, constraint (6.16) denotes the maximum number of links a child-UAV can support. Also, constraint (6.17) takes care of the minimum SINR criterion to maintain QoS. Furthermore, constraint (6.18) restricts an SCB to be connected with one particular child-UAV. Moreover, backhaul data rate limit, R_B, is represented in constraint (6.19).

It can be observed from the above optimization problem that for a fixed position of UAVs, the association of SCBs with UAVs is Binary Integer Linear Programming (BILP). To satisfy the constraints (6.15)–(6.19), the resultant optimization problem becomes NP-hard, and there is no standard method to solve such an optimization problem. In Section 6.7, we present an association algorithm, which is heuristic in nature, to associate the SCBs with UAVs.

6.7 Association of SCBs with UAVs

The association algorithm is addressed by keeping in view the distribution of SCBs and UAVs mentioned in Sections 6.5.1 and 6.5.2, respectively. In addition, the pseudo-code of the association algorithm is given in Algorithms 6.1 and 6.2. Mainly, the association algorithm is divided into two parts, i.e. Algorithm 6.1 will be executed at child-UAVs plus SCBs and Algorithm 6.2 at the parent-UAV. In the following, a brief summary of both the algorithms is described.

- In lines 1–3 of Algorithm 6.1, SCBs select the child-UAV, which gives the maximum SINR value. Mathematically, **A** contains a number of nonzero entries in each column and only one nonzero entry in each row. This shows that if the mth SCB is connected with the nth child-UAV, then there is value 1 in that particular position (i.e. mth row and nth column). Further, rest of the entries in the mth row are zero, which depicts that an SCB is only connected with one child-UAV, thus satisfying the constraint (6.18).

- In the next step, each child-UAV verifies the constraint (6.17) and it starts allocating the bandwidth[2] (based on demanded data rate, $r_{m,n}$, of SCB) by

---

2 Here, a cognitive radio framework can also be considered in our model so that additional bandwidth at the child-UAVs can efficiently be utilized [40].

Algorithm 6.1: Association of SCBs with UAVs

Input: $M, N, L^{max}, W, w_{m,n}, r_{m,n}, \Gamma_{m,n}$
Output: $\mathbf{A}$
// Initialize matrix A with all zeros
Initialize: $\mathbf{A} = \varnothing$
// Task at each SCB
1 **for** $m = 1$ **to** M **do**
2   Select child-UAV from which an mth SCB receives maximum SINR, Γ^{max}.
3   Update $a_{m,n} = 1$
    // Task at each child-UAV
4 **for** $n = 1$ **to** N **do**
5   If m^{th} SCB does not satisfy the constraint (6.17), then update $a_{m,n} = 0$
6   **Initialize counters**: $C_\ell = 0, C_w = 0$
    while $C_\ell < L_n^{max} \wedge C_w < W_n$ **do**
7   Find highest spectral efficient SCB
8   **if** $C_w + w_{m,n} \leq W_n$ **then**
9   Update $C_\ell = C_\ell + 1$ and $C_w = C_w + w_{m,n}$
10   **else**
11   Update $a_{m,n} = 0,$

Algorithm 6.2: Algorithm for Constraint (6.19)

// Task at parent-UAV
1 If $S_r < R_B$ then algorithm completes. Otherwise, following step is performed at the parent-UAV.
Initialize: S_r as total sum-rate of associated SCBs
2 **while** $S_r > R_B$ **do**
3   Select child-UAV with max. associated SCBs (to provide fairness)
    ▷Select SCB with min. data rate **min** $(r_{m,n})$ demand
4   De-associate the selected pair (m^{th} SCB of nth child-UAV) and update $a_{m,n} = 0, C_\ell = C_\ell - 1, S_r = S_r - r_{m,n}$ and $C_w = C_w - w_{m,n}$

---

keeping in view the constraint (6.15) and (6.16). However, each child-UAV serves higher spectral efficient, Ψ_{SE}, SCBs first, and is defined as

$$\Psi_{SE} = \frac{r_{m,n}}{w_{m,n}}. \tag{6.21}$$

The pseudo-code of this step is summarized in lines 4–11 of Algorithm 6.1.

- In the last step, every child-UAV forwards the aggregated information (i.e. SINR, demanded rate, etc.) to the parent-UAV, where the parent-UAV checks the backhaul link capacity by considering constraint (6.19). The pseudo-code (to verify backhaul link capacity) is summarized in Algorithm 6.2.

6.8 Results and Discussions

In this section, we first describe the simulation environment. Later on, a detailed analysis of the obtained results is presented.

6.8.1 Analysis of Results

We consider a dense urban environment where SCBs are distributed using *Matern type-I* hard-core process with a density of ρ per meter square. In addition, random data rate demands are assigned to deployed SCBs from a predefined data rate vector $\mathbf{r} = \{20, 40, 60, 80, 100\}$, where the values are in Mbps. Additionally, to satisfy the requirements of a 5G network, a minimum distance (d_{min}) of 250 m is maintained between the two SCBs [41]. Furthermore, the rest of the simulation parameters (unless stated specifically) and their description is given in Table 6.2. Moreover, the results are averaged over 1000 Monte Carlo realizations.

Table 6.2 Simulation parameters [2, 39].

Parameter	Value	Description
λ_c	0.15 m	Ratio: speed of light/frequency
a, b	9.61, 0.16	Environment constants
ϵ^ℓ, ϵ	$\{1, 20\}$ dB	Attenuation of LoS & NLoS paths
ρ	$2 \times 10^{-6} / \text{m}^2$	Density of SCBs
γ	2	Path loss exponent
S	16 km^2	Total area
P^t	1.5 W	Transmission power
σ^2_{noise}	-125 dB	Noise power
L^{max}	7	Maximum number of links
W	200 MHz	Maximum bandwidth
Γ^{min}	-5 dB	Minimum SINR
R_B	1.66 Gbps	Backhaul data rate

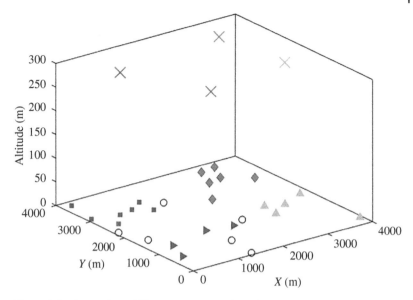

Figure 6.4 A snapshot of 3D placement of child-UAVs and the association of SCBs. × sign shows child-UAVs; unfilled markers denote unassociated SCBs; filled markers are associated SCBs with the respective × child-UAV.

Figure 6.4 represents a snapshot of the association problem. There are two network entities depicted in the figure, i.e. SCBs and the child-UAVs. All the child-UAVs, which are distributed using (6.2), are hovering at the same height, i.e. 300 m. Additionally, filled markers are the associated SCBs with the child-UAVs. Besides, unfilled circles are the unassociated SCBs. For example, the cluster indicated with square markers, all the SCBs are served by the respective child-UAV; however, there are a few unassociated ones. This is because child-UAV has reached to its maximum link limit, i.e. constraint (6.16); therefore, it cannot serve those SCBs. Importantly, the unassociated SCBs have also not been served by its neighboring child-UAVs, which can be because the SINR constraint is not satisfied, etc. Considering this position of SCBs and the child-UAVs, in the following, we evaluate the performance of the association algorithm under different constraints.

Figure 6.5 shows how the sum-rate of the overall network increases with the increase in available bandwidth (constraint (6.15)) and the number of links (constraint (6.16)), at each child-UAV. The trend portrays that the sum rate of the system increases exponentially with the increase in constraint (6.15) and constraint (6.16). Specifically, when $L^{max} = 1$, then there is no improvement in the sum rate when the bandwidth is increased, which is because the child-UAV cannot associate with more SCBs due to

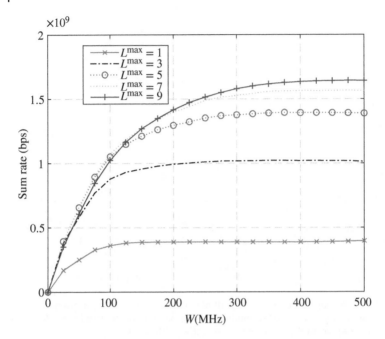

Figure 6.5 Comparison of sum rate by varying the constraint (6.15) and (6.16).

constraint (6.16). However, increasing L^{max} results in serving more SCBs; thus, the sum rate improves. In particular, it can be observed that with $L^{\mathrm{max}} = 9$, the maximum sum rate of the system has been achieved and we cannot go beyond as all the SCBs have been served till this point and there are no further data rates to add. Therefore, it can be summarized that increase in the sum rate is dependent on both, i.e. the number of available links at each child-UAV and the bandwidth resource.

Figure 6.6 reveals the behavior of the sum rate and the corresponding percentage of unassociated SCBs. It can be depicted that when the value of constraint (6.15) increases then the sum rate of the system increases, and the percentage of unassociated SCBs decreases. Nevertheless, it can be noted that all the SCBs are not served despite increasing the bandwidth and having a high value of constraint (6.16). This is because of constraint (6.19), which is limited to 1.5 Gbps. On the other side, the sum rate has been maximized (1.5 Gbps) by keeping in view the backhaul data rate limit.

Figure 6.7 depicts the performance improvement of the sum rate when the available number of links and the bandwidth resource are varied at each child-UAV. It can be concluded that the sum rate of the overall network increases with increase in both the constraints, i.e. constraint (6.15), and

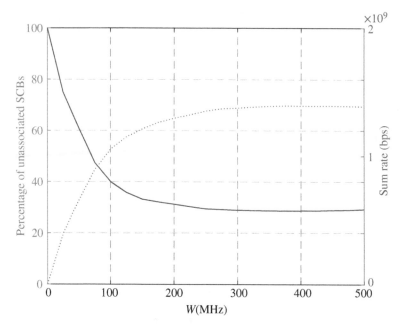

Figure 6.6 A comparison of unassociated SCBs and the sum rate with constraint (6.15) when $L^{max} = 10$ and $R_B = 1.5$ *Gbps*.

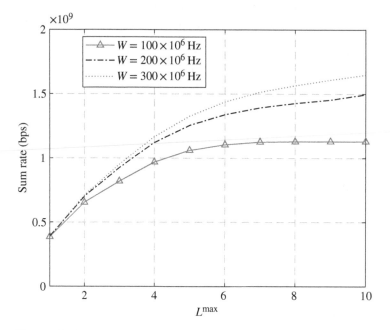

Figure 6.7 Performance of sum rate by varying the constraint (6.16) and having $R_B = 1.66$ Gbps.

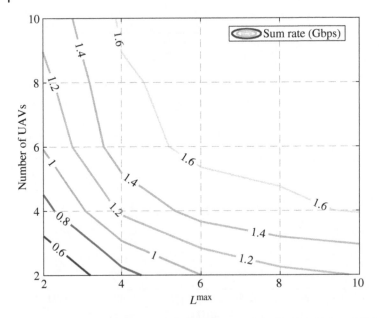

Figure 6.8 Sum rate's performance when the number of child-UAVs and L^{max} are varied; $R_B = 1.66$ Gbps, and $W = 300$ MHz.

(6.16). Nonetheless, it is important to note that in the case of the dotted curve, the sum rate has almost reached its maximum backhaul data rate limit, i.e. $R_B = 1.66$ Gbps.

Figure 6.8 portrays a contour map, where the performance of sum rate can be observed as the number of child-UAVs and the available links at each child-UAV are varied. It can be observed that the lower number of child-UAVs and lower value of the number of links bring fewer benefits. Thus, a higher number of child-UAVs and a higher value of L^{max} result in increasing the sum rate of the overall network.

In a nutshell, the above results demonstrate the performance of the association of SCBs with the UAVs by considering the various communication-related constraints. Favorable performance of the association algorithms (Algorithms 6.1 and 6.2) and the placement of UAVs have been observed.

6.9 Conclusion

This chapter focused on the vital role of UAVs in the mmWave and THz communication environment. In particular, the significance of UAVs deployment, and the benefits of mmWave and THz communication are addressed. Besides, the potential challenges of UAV communications are explained. In addition, the role of UAV communications in the fronthaul network is demonstrated. Specifically, the association and placement problem of UAVs is addressed by considering several communication-related constraints, e.g. bandwidth, SINR, and data rate. Additionally, an algorithm to deal with the backhaul link capacity is provided to enhance fairness in the overall network. Moreover, a detailed analysis of the results has been presented, which shows the significant performance of the addressed algorithms.

Bibliography

1 Xia, W., Semkin, V., Mezzavilla, M. et al. (2020). Multi-array designs for mmWave and sub-THz communication to UAVs. *2020 IEEE 21st International Workshop on Signal Processing Advances in Wireless Communications (SPAWC)*, IEEE, pp. 1–5.

2 Cheema, M.A., Shehzad, M.K., Qureshi, H.K. et al. (2020). A drone-aided blockchain-based smart vehicular network. *IEEE Transactions on Intelligent Transportation Systems* 1–11. https://doi.org/10.1109/TITS.2020.3019246.

3 Muntaha, S.T., Hassan, S.A., Jung, H., and Hossain, M.S. (2020). Energy efficiency and hover time optimization in UAV-based HetNets. *IEEE Transactions on Intelligent Transportation Systems* 1–9. https://doi.org/10.1109/TITS.2020.3015256.

4 Naqvi, S.A.R., Hassan, S.A., Pervaiz, H., and Ni, Q. (2018). Drone-aided communication as a key enabler for 5G and resilient public safety networks. *IEEE Communications Magazine* 56 (1): 36–42.

5 Xiao, Z., Zhu, L., and Xia, X.-G. (2020). UAV communications with millimeter-wave beamforming: potentials, scenarios, and challenges. *China Communications* 17 (9): 147–166.

6 Polese, M., Bertizzolo, L., Bonati, L. et al. (2020). An experimental mmWave channel model for UAV-to-UAV communications. *Proceedings of the 4th ACM Workshop on Millimeter-Wave Networks and Sensing Systems*, pp. 1–6.

7 Jan, M.A., Hassan, S.A., and Jung, H. (2019). QoS-based performance analysis of mmWave UAV-assisted 5G hybrid heterogeneous network. *2019 IEEE Global Communications Conference (GLOBECOM)*, IEEE, pp. 1–6.

8 Shehzad, M.K., Hassan, S.A., Luque-Nieto, M.A. et al. (2020). Energy efficient placement of UAVs in wireless backhaul networks. *Proceedings of the 2nd ACM MobiCom Workshop on Drone Assisted Wireless Communications for 5G and Beyond*, pp. 1–6.

9 Gapeyenko, M., Petrov, V., Moltchanov, D. et al. (2018). Flexible and reliable UAV-assisted backhaul operation in 5G mmWave cellular networks. *IEEE Journal on Selected Areas in Communications* 36 (11): 2486–2496.

10 Kim, J. and Kim, Y. (2008). Moving ground target tracking in dense obstacle areas using UAVs. *IFAC Proceedings Volumes* 41 (2): 8552–8557.

11 Chen, C., Grier, A., Malfa, M. et al. (2017). High-speed optical links for UAV applications. *Free-Space Laser Communication and Atmospheric Propagation XXIX*, Volume 10096, International Society for Optics and Photonics, p. 1009615.

12 Sarieddeen, H., Saeed, N., Al-Naffouri, T.Y., and Alouini, M.-S. (2020). Next generation terahertz communications: a rendezvous of sensing, imaging, and localization. *IEEE Communications Magazine* 58 (5): 69–75.

13 Zia-ul-Mustafa, R. and Hassan, S.A. (2019). Machine learning-based context aware sequential initial access in 5G mmWave systems. *2019 IEEE Globecom Workshops (GC Wkshps)*, IEEE, pp. 1–6.

14 Guan, Z. and Kulkarni, T. (2019). On the effects of mobility uncertainties on wireless communications between flying drones in the mmWave/THz bands. *IEEE INFOCOM 2019-IEEE Conference on Computer Communications Workshops (INFOCOM WKSHPS)*, IEEE, pp. 768–773.

15 Dabiri, M.T., Safi, H., Parsaeefard, S., and Saad, W. (2020). Analytical channel models for millimeter wave UAV networks under hovering fluctuations. *IEEE Transactions on Wireless Communications* 19 (4): 2868–2883.

16 Brighente, A., Cerutti, M., Nicoli, M. et al. (2020). Estimation of wideband dynamic mmWave and THz channels for 5G systems and beyond. *IEEE Journal on Selected Areas in Communications* 38 (9): 2026–2040.

17 Xing, Y., Kanhere, O., Ju, S., and Rappaport, T.S. (2019). Indoor wireless channel properties at millimeter wave and sub-terahertz frequencies. *2019 IEEE Global Communications Conference (GLOBECOM)*, IEEE, pp. 1–6.

18 Elayan, H., Amin, O., Shubair, R.M., and Alouini, M.-S. (2018). Terahertz communication: the opportunities of wireless technology beyond 5G. *2018 International Conference on Advanced Communication Technologies and Networking (CommNet)*, IEEE, pp. 1–5.

19 Raja, A.A., Jamshed, M.A., Pervaiz, H., and Hassan, S.A. (2020). Performance analysis of UAV-assisted backhaul solutions in THz enabled hybrid heterogeneous network. *IEEE INFOCOM 2020-IEEE Conference* *on Computer Communications Workshops (INFOCOM WKSHPS)*, IEEE, pp. 628–633.

20 Chen, Z., Ma, X., Zhang, B. et al. (2019). A survey on terahertz communications. *China Communications* 16 (2): 1–35.

21 Saeed, A., Gurbuz, O., and Akkas, M.A. (2020). Terahertz communications at various atmospheric altitudes. *Physical Communication* 41 101113.

22 Sambo, Y.A., Klaine, P.V., Nadas, J.P.B., and Imran, M.A. (2019). Energy minimization UAV trajectory design for delay-tolerant emergency communication. *2019 IEEE International Conference on Communications Workshops (ICC Workshops)*, IEEE, pp. 1–6.

23 Mozaffari, M., Saad, W., Bennis, M. et al. (2019). A tutorial on UAVs for wireless networks: applications, challenges, and open problems. *IEEE Communications Surveys & Tutorials* 21 (3): 2334–2360.

24 Shehzad, M.K. (2019). Association of small cell base stations with UAVs in HetNets. PhD thesis. Association of Small Cell Base Stations with UAVs in HetNets By Muhammad ....

25 Amer, R., Saad, W., ElSawy, H. et al. (2019). Caching to the sky: performance analysis of cache-assisted CoMP for cellular-connected UAVs. *2019 IEEE Wireless Communications and Networking Conference (WCNC)*, IEEE, pp. 1–6.

26 Shehzad, M.K., Hassan, S.A., Mahmood, A., and Gidlund, M. (2019). On the association of small cell base stations with UAVs using unsupervised learning. *IEEE Vehicular Technology Conference (VTC-Spring)*, May 2019.

27 Alzenad, M., Shakir, M.Z., Yanikomeroglu, H., and Alouini, M.-S. (2018). FSO-based vertical backhaul/fronthaul framework for 5G+ wireless networks. *IEEE Communications Magazine* 56 (1): 218–224.

28 Leitgeb, E., Zettl, K., Muhammad, S.S. et al. (2007). Investigation in free space optical communication links between unmanned aerial vehicles (UAVs). *2007 9th International Conference on Transparent Optical Networks*, Volume 3, IEEE, pp. 152–155.

29 Najafi, M., Ajam, H., Jamali, V. et al. (2018). Statistical modeling of FSO fronthaul channel for drone-based networks. *2018 IEEE International Conference on Communications (ICC)*, IEEE, pp. 1–7.

30 Matérn, B. (1986). *Spatial Variation, Springer Lecture Notes in Statistics*, 36. Springer.

31 Bor-Yaliniz, R.I., El-Keyi, A., and Yanikomeroglu, H. (2016). Efficient 3-D placement of an aerial base station in next generation cellular networks. *2016 IEEE International Conference on Communications (ICC)*, IEEE, pp. 1–5.

32 Kalantari, E., Shakir, M.Z., Yanikomeroglu, H., and Yongacoglu, A. (2017). Backhaul-aware robust 3D drone placement in 5G+ wireless networks. *2017 IEEE International Conference on Communications Workshops (ICC Workshops)*, IEEE, pp. 109–114.

33 Alzenad, M., El-Keyi, A., Lagum, F., and Yanikomeroglu, H. (2017). 3-D placement of an unmanned aerial vehicle base station (UAV-BS) for energy-efficient maximal coverage. *IEEE Wireless Communications Letters* 6 (4): 434–437.

34 Kalantari, E., Yanikomeroglu, H., and Yongacoglu, A. (2016). On the number and 3D placement of drone base stations in wireless cellular networks. *2016 IEEE 84th Vehicular Technology Conference (VTC-Fall)*, IEEE, pp. 1–6.

35 Mozaffari, M., Saad, W., Bennis, M., and Debbah, M. (2016). Efficient deployment of multiple unmanned aerial vehicles for optimal wireless coverage. *IEEE Communications Letters* 20 (8): 1647–1650.

36 Sharma, V., Srinivasan, K., Chao, H.-C. et al. (2017). Intelligent deployment of UAVs in 5G heterogeneous communication environment for improved coverage. *Journal of Network and Computer Applications* 85: 94–105.

37 Galkin, B., Kibilda, J., and DaSilva, L.A. (2016). Deployment of UAV-mounted access points according to spatial user locations in two-tier cellular networks. *2016 Wireless Days (WD)*, IEEE, pp. 1–6.

38 Likas, A., Vlassis, N., and Verbeek, J.J. (2003). The global k-means clustering algorithm. *Pattern Recognition* 36 (2): 451–461.

39 Al-Hourani, A., Kandeepan, S., and Lardner, S. (2014). Optimal LAP altitude for maximum coverage. *IEEE Wireless Communications Letters* 3 (6): 569–572.

40 Mitola, J. (1999). Cognitive radio for flexible mobile multimedia communications. *1999 IEEE International Workshop on Mobile Multimedia Communications (MoMuC'99)(Cat. No. 99EX384)*, IEEE, pp. 3–10.

41 Nordrum, A., Clark, K. IEEE Spectrum Staff (2017). *5G Bytes: Small Cells Explained*. New York: IEEE Spectrum. https://spectrum.ieee.org/video/telecom/wireless/5g-bytes-small-cells-explained (accessed 31 March 2021).

7

UAV-Enabled Cooperative Jamming for Physical Layer Security in Cognitive Radio Network

Phu X. Nguyen[1], Hieu V. Nguyen[2], Van-Dinh Nguyen[3], and Oh-Soon Shin[4]

[1]Department of Computer Fundamentals, FPT University, Ho Chi Minh City, Vietnam
[2]The University of Danang - Advanced Institute of Science and Technology, Da Nang, Vietnam
[3]Interdisciplinary Centre for Security, Reliability and Trust (SnT), University of Luxembourg, Luxembourg
[4]School of Electronic Engineering, Soongsil University, Seoul, South Korea

7.1 Introduction

Ericsson reports that 5G subscriptions and the number of IoT connections are estimated to reach 2.8 and 25 billion, respectively, globally by the end of 2025 [1, 2]. The unprecedented increase in communications is becoming a huge challenge due to limited spectral resources at terrestrial communication networks. This raises concerns on how to assist the communication needs of a massive number of users with restricted resources without affecting the network performance. Cognitive radio networks (CRNs) have been introduced as a potential technology to improve spectral efficiency [3–7]. The main idea of CRNs is to allow secondary users (or unlicensed users) to access the underutilized spectrum while keeping restricted interference to the primary users (or licensed users) [7–9].

Besides many advantages, cognitive radio communications still have limitations that need to be overcome. Specifically, due to the accessibility of the frequency bands and broadcast nature of wireless communications, confidential communication between the secondary users can be overheard and decoded by the eavesdropper (Eve) [10–12]. Complexity-based cryptography can be effective when the computational resources of Eves are too limited to decipher the secret key [13]. However, Eve's computing power has been significantly evolving, while a trust infrastructure for ensuring confidential communications is expensive to deploy. Physical

Autonomous Airborne Wireless Networks, First Edition.
Edited by Muhammad Ali Imran, Oluwakayode Onireti, Shuja Ansari, and Qammer H. Abbasi.
© 2021 John Wiley & Sons Ltd. Published 2021 by John Wiley & Sons Ltd.

layer security (PLS) appears as an alternative security paradigm due to the security obtained in wireless communications without the secret key management costs [14, 15]. This topic has attracted significant attention from researchers, in which cooperative jamming schemes are one of the techniques commonly used to degrade the decoding capability of the Eve [16–18]. Nevertheless, most of the conventional cooperative jamming schemes are employed based on ground jammers with fixed locations and limited antenna heights. This leads to three major challenges. First, when jammers are deployed far away from the Eves, the effect of jamming noise (JN) is significantly decreased, and thus the security performance degrades. Second, the legitimate transmitter needs to know the channel state information (CSI) of the Eves to perform jamming. This is a big challenge since the Eves are usually passive devices for which it is difficult to estimate the CSI. Finally, in CRNs, the secrecy rate enhancement of the secondary system may also influence the primary system since the interference powers (i.e. transmit power of secondary transmitter [ST] and jamming power of UAV) to primary receiver (PR) may exceed the predefined threshold.

Recently, unmanned aerial vehicles (UAVs) play a key role in various sectors and will enable a myriad of applications such as aerial surveillance, traffic control, emergency communications, and package delivery [19–23]. Regarding the PLS, a UAV can easily move to an optimal location (e.g. a few hundred meters from the ground) with low blockage to assist any ground users by sending a friendly jammer. In [24], the PLS for transmission in underlay CRNs was studied by examining a system including a primary transmitter–receiver pair and a secondary transmitter–receiver pair in the presence of an Eve. Specifically, primary source (PS) transmits JN, which is combined with the expected signal, to interfere with Eve. However, when Eves are distributed far away from the PS, the effect of JN is significantly reduced, and thus the secrecy rate deteriorates. The authors in [14] showed that if the power of the jammer was sufficiently high, the legitimate transmissions between cognitive users and cognitive sources can be destroyed due to the broadcast nature of radio propagation. Therefore, it was necessary to adjust the power of the jammer to meet the quality of legitimate transmissions in CRNs. UAV-enabled mobile relaying, which aimed to improve the secrecy rate of ground users, was considered in [25]. The authors in [26] proposed UAV-enabled JN to degrade the quality of the ground wiretap channel. In [27], the transmit power and UAV trajectory were jointly optimized to maximize the secrecy rate of the air-to-ground communication link, in which the UAV plays a role as a transmitter. The UAV-aided secure communications with a cooperative jamming UAV, where the minimum secrecy rate was maximized by jointly optimizing the transmit power and UAV trajectory as well as the user scheduling, was studied in [28].

Different from [25–28], our work investigates PLS for a CRN, in which a UAV is introduced as a friendly jammer to assist the ground cognitive users. We propose an efficient algorithm, where the transmission power at the transmitters and the UAV's trajectory are jointly optimized to solve the problem of achieving the secrecy rate maximization (SRM). To the best of our knowledge, *this is the first work that considers UAV-enabled cooperative jamming to improve the secrecy rate of CRN*. Hence, our main contributions can be summarized as follows:

- Unlike the conventional jamming schemes that employ ground jammers with fixed locations, a mobile UAV acts as a friendly jammer to transmit JN effectively by flying closer/farther to ground nodes. This leads to a considerable increase in the average secrecy rate but still satisfies the condition of the underlay cognitive radio.
- The original optimization problem is transformed into a more tractable design, which is a suboptimal but safe solution. Then, an efficient iterative algorithm based on the inner approximation (IA) method is proposed to tackle the non-convex problem effectively with a fast convergence rate.

7.2 System Model

This research mainly focuses on maximizing the average secrecy rate of the secondary system while guaranteeing that the transmit powers do not affect the quality of service (QoS) of the primary system.

7.2.1 Signal Model

We study a CRN on the ground composed of one secondary transmitter (ST) and one secondary receiver (SR) in the presence of one PR, as illustrated in Figure 7.1. In the secondary network, one Eve attempts to overhear and decode confidential messages from the ST to SR. To improve the PHY layer security of CRN, we introduce one UAV to transmit the friendly JN to degrade the decoding capability of Eve. Here, the coordinates of ST, SR, PR, and Eve are denoted by $(0, 0, 0)$, $(x_S, y_S, 0)$, $(x_P, y_P, 0)$, and $(x_E, y_E, 0)$, respectively. In this work, T is discretized into N equal time slots, where each time slot is given by $\delta_t = \frac{T}{N}$. The number of time slots must be sufficiently large to guarantee that a UAV's location is almost unchanged within each time slot. Therefore, the time-varying horizontal coordinate

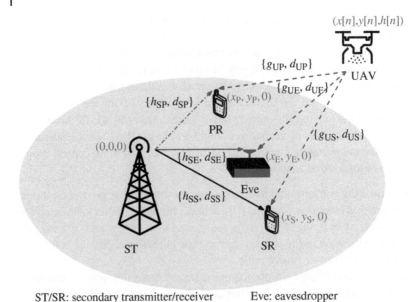

ST/SR: secondary transmitter/receiver   Eve: eavesdropper
PR: primary receiver   UAV: unmanned aerial vehicle

Figure 7.1 The UAV-enabled cooperative jamming in cognitive radio system.

can be defined by $\mathbf{q}[n] = (x[n], y[n], H)$, $n \in \mathcal{N} \triangleq \{1, \dots, N\}$. The initial and final locations of UAV are given as (x_0, y_0, H) and (x_f, y_f, H), respectively. The UAV's trajectory must satisfy the maximum speed constraint $\|\mathbf{q}'_m(t)\| \leq V_{\max}$, $0 \leq t \leq T$, where $\mathbf{q}'_m(t)$ denotes the derivative of the UAV's location with respect to time t, and $V_{\max}$ is the maximum speed of the UAV. As a result, the mobility constraints of the UAV can be expressed as

$$(x[1] - x_0)^2 + (y[1] - y_0)^2 \leq L_{\max}^2, \tag{7.1a}$$

$$(x[n + 1] - x[n])^2 + (y[n + 1] - y[n])^2 \leq L_{\max}^2, \ \forall n \in \mathcal{N}, \tag{7.1b}$$

$$(x_f - x[N])^2 + (y_f - y[N])^2 = 0, \tag{7.1c}$$

where $L_{\max} = V_{\max} \delta_t$ denotes the maximum distance that the UAV can move within each time slot.

We assume that the air-to-ground channels are the line-of-sight (LoS) links. Thus, the channel power gain from the UAV to the ground nodes (SR, PR, Eve) at time slot n can be expressed by the free space path loss as $g_{U_i}[n] = \rho_0 d_{U_i}^{-2}[n]$, $i \in \{S, P, E\}$:

$$g_{\mathrm{US}}[n] = \rho_0 d_{\mathrm{US}}^{-2}[n] = \frac{\rho_0}{\left(x_S - x[n]\right)^2 + \left(y_S - y[n]\right)^2 + H^2}, \quad \forall n \in \mathcal{N},$$

(7.2a)

$$g_{\mathrm{UP}}[n] = \rho_0 d_{\mathrm{UP}}^{-2}[n] = \frac{\rho_0}{\left(x_P - x[n]\right)^2 + \left(y_P - y[n]\right)^2 + H^2}, \quad \forall n \in \mathcal{N},$$

(7.2b)

$$g_{\mathrm{UE}}[n] = \rho_0 d_{\mathrm{UE}}^{-2}[n] = \frac{\rho_0}{\left(x_E - x[n]\right)^2 + \left(y_E - y[n]\right)^2 + H^2}, \quad \forall n \in \mathcal{N},$$

(7.2c)

where ρ_0 is the channel power gain at the reference distance $d_0 = 1$ m, while d_{U_i} is the distance between UAV and ground nodes (SR, PR, Eve). Let $P_U[n]$ denote the transmit power at the UAV in time slot n. In practice, it is essential to consider both average and peak powers. Thus, the transmit power constraints are expressed as

$$\frac{1}{N} \sum_{n \in \mathcal{N}} P_U[n] \leq \overline{P_U},$$

(7.3a)

$$0 \leq P_U[n] \leq P_U^{\max}, \quad \forall n \in \mathcal{N}.$$

(7.3b)

We next consider the ground-to-ground channels, which are assumed to experience independent Rayleigh fading, and thus the power channel gain from cognitive source ST to SR, PR, and Eve, denoted by h_{SS}, h_{SP}, and h_{SE}, can be expressed as

$$h_{\mathrm{SS}} = \rho_0 d_{\mathrm{SS}}^{-\varphi} \xi_{\mathrm{SS}},$$

(7.4a)

$$h_{\mathrm{SP}} = \rho_0 d_{\mathrm{SP}}^{-\varphi} \xi_{\mathrm{SP}},$$

(7.4b)

$$h_{\mathrm{SE}} = \rho_0 d_{\mathrm{SE}}^{-\varphi} \xi_{\mathrm{SE}},$$

(7.4c)

where d_{SS}, d_{SP}, and d_{SE} denote the distance between ST and the ground nodes (SR, PR, Eve), φ denotes the path loss exponent, and ξ_i, $i \in \{\mathrm{SS}, \mathrm{SP}, \mathrm{SE}\}$, is an exponentially distributed random variable with unit mean. Similarly to (7.3), the transmit power constraints in the ground-to-ground channel are shown as

$$\frac{1}{N} \sum_{n \in \mathcal{N}} P_S[N] \leq \overline{P_S},$$

(7.5a)

$$0 \leq P_S[n] \leq P_S^{\max}, \quad \forall n \in \mathcal{N}.$$

(7.5b)

The average achievable rates from the ST to SR and to Eve at time slot n can be given by

$$R_S[n] = \mathbb{E}_{h_{SS}}\left\{\log_2\left(1 + \frac{P_S[n]h_{SS}}{P_U[n]g_{US}[n] + \sigma^2}\right)\right\}, \quad \forall n \in \mathcal{N}, \quad (7.6a)$$

$$R_E[n] = \mathbb{E}_{h_{SE}}\left\{\log_2\left(1 + \frac{P_S[n]h_{SE}}{P_U[n]g_{UE}[n] + \sigma^2}\right)\right\}, \quad \forall n \in \mathcal{N}, \quad (7.6b)$$

where $\mathbb{E}\{.\}$ describes the expectation operator with respect to ground fading channels and σ^2 denotes the additive white Gaussian noise (AWGN) power. Thus, the average achievable secrecy rate of the secondary system over the total N time slots can be defined as [24]

$$R_{\text{sec}} = \frac{1}{N}\sum_{n \in \mathcal{N}}[R_S[n] - R_E[n]]^+, \quad (7.7)$$

where $[x]^+ \triangleq \max\{0, x\}$.

As mentioned before, to guarantee that the QoS of the primary system is not affected by interference powers caused by the UAV and ST, the received average interference power must satisfy the following constraint [25]:

$$\frac{1}{N}\sum_{n \in \mathcal{N}}\left(\mathbb{E}_{h_{SP}}\{P_S[n]h_{SP}\} + P_U[n]g_{UP}[n]\right) \leq \varepsilon. \quad (7.8)$$

Our objective is to maximize the average achievable secrecy rate R_{sec} of the secondary system by jointly optimizing the UAV's trajectory $\mathbf{q} \triangleq \{x[n], y[n], h[n]\}_{n \in \mathcal{N}}$ and transmit powers $\mathbf{p} \triangleq \{p_S[n], p_U[n]\}_{n \in \mathcal{N}}$ over the total N time slots while meeting the average interference power constraint at PR. From constraints (7.1), (7.3), (7.5), and (7.8), the optimization problem is stated as follows:

$$\text{P1:} \quad \max_{\mathbf{q},\mathbf{p}} R_{\text{sec}} \quad (7.9a)$$

$$\text{s.t.} \quad (x[1] - x_0)^2 + (y[1] - y_0)^2 \leq L_{\max}^2, \quad (7.9b)$$

$$(x[n+1] - x[n])^2 + (y[n+1] - y[n])^2 \leq L_{\max}^2, \quad \forall n \in \mathcal{N}, \quad (7.9c)$$

$$(x_f - x[N])^2 + (y_f - y[N])^2 = 0, \quad (7.9d)$$

$$\frac{1}{N}\sum_{n \in \mathcal{N}}P_U[n] \leq \overline{P_U}, \quad (7.9e)$$

$$0 \leq P_U[n] \leq P_U^{\max}, \quad \forall n \in \mathcal{N}, \quad (7.9f)$$

$$\frac{1}{N}\sum_{n \in \mathcal{N}}P_S[n] \leq \overline{P_S}, \quad (7.9g)$$

$$0 \leq P_S[n] \leq P_S^{\max}, \quad \forall n \in \mathcal{N}, \quad (7.9h)$$

$$\frac{1}{N}\sum_{n \in \mathcal{N}}\left(\mathbb{E}_{h_{SP}}\{P_S[n]h_{SP}\} + P_U[n]g_{UP}[n]\right) \leq \varepsilon. \quad (7.9i)$$

7.2.2 Optimization Problem Formulation

Optimization (7.9) is really a challenge due to two reasons. Firstly, the objective function (7.9a) and constraint (7.9i) are non-convex. Secondly, the objective function is non-smooth at point zero due to the operator $[.]^+$. To make the problem tractable, we first transform (7.9) into the equivalent problem as follows:

$$\text{P2: } \max_{\mathbf{q}, \mathbf{p}} \ R_{\text{sec}}^{\text{eq}} \tag{7.10a}$$

$$\text{s.t.} \quad (x[1] - x_0)^2 + (y[1] - y_0)^2 \le L_{\max}^2, \tag{7.10b}$$

$$(x[n+1] - x[n])^2 + (y[n+1] - y[n])^2 \le L_{\max}^2, \ \forall n \in \mathcal{N}, \tag{7.10c}$$

$$(x_f - x[N])^2 + (y_f - y[N])^2 = 0, \tag{7.10d}$$

$$\frac{1}{N} \sum_{n \in \mathcal{N}} P_{\text{U}}[n] \le \bar{P}_{\text{U}}, \tag{7.10e}$$

$$0 \le P_{\text{U}}[n] \le P_{\text{U}}^{\max}, \ \forall n \in \mathcal{N}, \tag{7.10f}$$

$$\frac{1}{N} \sum_{n \in \mathcal{N}} P_{\text{S}}[n] \le \bar{P}_{\text{S}}, \tag{7.10g}$$

$$0 \le P_{\text{S}}[n] \le P_{\text{S}}^{\max}, \ \forall n \in \mathcal{N}, \tag{7.10h}$$

$$\frac{1}{N} \sum_{n \in \mathcal{N}} \left(\mathbb{E}_{h_{\text{SP}}} \left\{ P_{\text{S}}[n] h_{\text{SP}} \right\} + P_{\text{U}}[n] g_{\text{UP}}[n] \right) \le \varepsilon, \tag{7.10i}$$

where

$$R_{\text{sec}}^{\text{eq}} = \frac{1}{N} \sum_{n \in \mathcal{N}} \left(R_{\text{S}}[n] - R_{\text{E}}[n] \right). \tag{7.11}$$

Intuitively, we can see that (7.10) has the same optimal solution as (7.9). Since the value of the objective function is less than zero, the objective function can be increased to zero by reducing the power of ST to zero [25].

7.3 Proposed Algorithm

Although the original optimization problem has been transformed into a more tractable form by removing the operator $[.]^+$, it is obvious that problem (7.10) is non-convex. To tackle this problem, we introduce an iterative algorithm based on the inner approximation (IA) method to transform (7.10) into a convex optimization problem and obtain an approximate solution.

7.3.1 Tractable Formulation for the Optimization Problem P2

In order to make P2 tractable, we attempt to remove the expectation operator in (7.10a) and (7.10i) as follows.

7.3.1.1 Tractable Formulation for $R_S[n]$

In this section, the expectation operator of $R_S[n]$ is removed by finding its lower bound.

$$R_S[n] = \mathbb{E}_{h_{ss}}\left\{\log_2\left(1 + \frac{P_S[n]h_{ss}}{P_U[n]g_{US}[n] + \sigma^2}\right)\right\}. \tag{7.12}$$

Let

$$X[n] = \frac{P_S[n]h_{ss}}{P_U[n]g_{US}[n] + \sigma^2}. \tag{7.13}$$

From (7.2a) and (7.4a), $X[n]$ can be rewritten as

$$X[n] = \frac{P_S[n]\rho_0 d_{ss}^{-\varphi}\xi_{ss}}{\dfrac{P_U[n]\rho_0}{\left(x_S - x[n]\right)^2 + \left(y_S - y[n]\right)^2 + H^2} + \sigma^2}. \tag{7.14}$$

$X[n]$ is an exponential distributed random variable with parameter

$$\lambda_X[n] = \frac{d_{ss}^{\varphi}}{\dfrac{\rho_0 P_S[n]}{P_U[n]g_{US}[n] + \sigma^2}}. \tag{7.15}$$

$R_S[n]$ can be reformulated as

$$R_S[n] = \mathbb{E}_{h_{ss}}\left\{\log_2\left(1 + X[n]\right)\right\} = \mathbb{E}_{h_{ss}}\left\{\log_2\left(1 + e^{\ln X[n]}\right)\right\}. \tag{7.16}$$

Since $\log_2\left(1 + e^x\right)$ is the convex function, by using Jensen's inequality, $R_S[n]$ is lower bounded as

$$R_S[n] = \mathbb{E}_{h_{ss}}\left\{\log_2\left(1 + e^{\ln X[n]}\right)\right\} \geq \log_2\left(1 + e^{\mathbb{E}_{h_{ss}}\{\ln X[n]\}}\right). \tag{7.17}$$

As mentioned before, $X[n]$ is an exponential distribution random variable with parameter $\lambda_X[n]$. Therefore, $\mathbb{E}_{h_{ss}}\{\ln X[n]\}$ is given by

$$\mathbb{E}_{h_{ss}}\{\ln X[n]\} = \int_0^\infty \ln x \lambda_X[n] e^{-\lambda_X[n]x}\,dy = -\ln \lambda_X[n] - k, \tag{7.18}$$

where k is the Euler constant. From (7.18), (7.17) is rewritten as

$$R_S[n] \geq R_S^{\text{LB}}[n] = \log_2\left(1 + A[n]\right), \tag{7.19}$$

where

$$A[n] = \frac{e^{-k}\gamma_0 P_S[n]d_{ss}^{-\varphi}}{\dfrac{\gamma_0 P_U[n]}{\left(x_U - x[n]\right)^2 + \left(y_U - y[n]\right)^2 + H^2} + 1},$$

$$\gamma_0 = \frac{\rho}{\sigma^2}. \tag{7.20}$$

7.3.1.2 Tractable Formulation for $R_E[n]$

We next find the upper bound of $R_E[n]$.

$$R_E[n] = \mathbb{E}_{h_{SE}} \left\{ \log_2 \left(1 + \frac{P_S[n]h_{SE}}{P_U[n]g_{UE}[n] + \sigma^2} \right) \right\}. \tag{7.21}$$

Let

$$Y[n] = \frac{P_S[n]h_{SE}}{P_U[n]g_{UE}[n] + \sigma^2}. \tag{7.22}$$

From (7.2c) and (7.4c), $Y[n]$ can be rewritten as

$$Y[n] = \frac{P_S[n]\rho_0 d_{SE}^{-\varphi}\xi_{SE}}{\dfrac{P_U[n]\rho_0}{\left(x_E - x[n]\right)^2 + \left(y_E - y[n]\right)^2 + H^2} + \sigma^2}. \tag{7.23}$$

Similarly to $X[n]$, $Y[n]$ is also an exponential distribution random variable with parameter

$$\lambda_Y[n] = \frac{d_{SE}^{\varphi}}{\dfrac{\rho_0 P_S[n]}{P_U[n]g_{UE}[n] + \sigma^2}}. \tag{7.24}$$

Since $\log_2(1 + x)$ is a concave function, we also apply Jensen's inequality to find the upper bound of $R_E[n]$:

$$R_E[n] = \mathbb{E}_{h_{SE}} \{\log_2(1 + Y[n])\}$$

$$\leq \log_2 \left(1 + \mathbb{E}_{h_{SE}} \{Y[n]\} \right) = \log_2 \left(1 + \frac{1}{\lambda_Y[n]} \right)$$

$$= R_E^{UB}[n] = \log_2(1 + B[n]), \tag{7.25}$$

where

$$B[n] = \frac{e^{-k}\gamma_0 P_S[n]d_{SE}^{-\varphi}}{\dfrac{\gamma_0 P_U[n]}{\left(x_E - x[n]\right)^2 + \left(y_E - y[n]\right)^2 + H^2} + 1}, \tag{7.26}$$

$$\gamma_0 = \frac{\rho}{\sigma^2}.$$

7.3.1.3 Tractable Formulation for Constraint (7.10i)

The expectation operator in (7.10i) is removed as

$$\mathbb{E}_{h_{SP}} \{P_S[n]h_{SP}\} = \mathbb{E}_{h_{SP}} \{\rho_0 d_{SP}^{-\varphi}P_S[n]\xi_{SP}\}. \tag{7.27}$$

Since $\rho_0 d_{SP}^{-\varphi}P_S[n]\xi_{SP}$ is also an exponential distribution random variable with $\lambda[n] = \frac{1}{\rho_0 P_S[n]}d_{SP}^{\varphi}$, the expectation operator $\mathbb{E}_{h_{SP}}\{\rho_0 d_{SP}^{-\varphi}P_S[n]\xi_{SP}\}$ is given as

$$\mathbb{E}_{h_{SP}} \{\rho_0 d_{SP}^{-\varphi}P_S[n]\xi_{SP}\} = \rho_0 d_{SP}^{-\varphi}P_S[n]. \tag{7.28}$$

Thus, constraint (7.8) can be rewritten as

$$\frac{1}{N} \sum_{n \in \mathcal{N}} \left(\rho_0 d_{\text{SP}}^{-\varphi} P_{\text{S}}[n] + P_{\text{U}}[n] g_{\text{UP}}[n] \right) \leq \varepsilon. \tag{7.29}$$

7.3.1.4 Safe Optimization Problem

Using (7.19), (7.25), and (7.29), the optimization problem (7.10) can be transformed into a safe design:

$$\text{P3:} \quad \max_{\mathbf{q}, \mathbf{p}} \; R_{\text{sec}}^{\text{safe}} \triangleq \frac{1}{N} \sum_{n \in \mathcal{N}} \left(R_{\text{S}}^{\text{LB}}[n] - R_{\text{E}}^{\text{UB}}[n] \right) \tag{7.30a}$$

$$\text{s.t.} \quad (x[1] - x_0)^2 + (y[1] - y_0)^2 \leq L_{\max}^2, \tag{7.30b}$$

$$(x[n+1] - x[n])^2 + (y[n+1] - y[n])^2 \leq L_{\max}^2, \; \forall n \in \mathcal{N}, \tag{7.30c}$$

$$(x_f - x[N])^2 + (y_f - y[N])^2 = 0, \tag{7.30d}$$

$$\frac{1}{N} \sum_{n \in \mathcal{N}} P_{\text{U}}[n] \leq \bar{P}_{\text{U}}, \tag{7.30e}$$

$$0 \leq P_{\text{U}}[n] \leq P_{\text{U}}^{\max}, \; \forall n \in \mathcal{N}, \tag{7.30f}$$

$$\frac{1}{N} \sum_{n \in \mathcal{N}} P_{\text{S}}[n] \leq \bar{P}_{\text{S}}, \tag{7.30g}$$

$$0 \leq P_{\text{S}}[n] \leq P_{\text{S}}^{\max}, \; \forall n \in \mathcal{N}, \tag{7.30h}$$

$$\frac{1}{N} \sum_{n \in \mathcal{N}} \left(\rho_0 d_{\text{SP}}^{-\varphi} P_{\text{S}}[n] + P_{\text{U}}[n] g_{\text{UP}}[n] \right) \leq \varepsilon. \tag{7.30i}$$

Remark 7.1 We notice that problem (7.30) is a safe design to problem (7.10) in the sense that any feasible point of the former is also feasible to the latter, but not vice versa due to (7.19) and (7.25). As a result, problem (7.30) provides a lower bound objective compared to (7.10), i.e. $R_{\text{sec}}^{\text{eq}} \geq R_{\text{sec}}^{\text{safe}}$. Thus, in what follows, we will consider the safe optimization problem of (7.30) in the sequel of the chapter, instead of (7.10).

7.3.2 Proposed IA-Based Algorithm

The IA method [29] is utilized to solve the optimization problem (7.30) in a single layer. By introducing slack variables $\mathbf{r} \triangleq \{r_{\text{S}}[n], r_{\text{E}}[n]\}_{n \in \mathcal{N}}$, we transform (7.30) into an equivalent design as follows:

$$\text{P4: } \max_{q,p,r} R_{\text{sec}}^{\text{LB}} \triangleq \frac{1}{N} \sum_{n \in \mathcal{N}} \left(r_{\text{S}}[n] - r_{\text{E}}[n] \right) \tag{7.31a}$$

$$\text{s.t. } (x[1] - x_0)^2 + (y[1] - y_0)^2 \le L_{\max}^2, \tag{7.31b}$$

$$(x[n+1] - x[n])^2 + (y[n+1] - y[n])^2 \le L_{\max}^2, \ \forall n \in \mathcal{N}, \tag{7.31c}$$

$$(x_f - x[N])^2 + (y_f - y[N])^2 = 0, \tag{7.31d}$$

$$\frac{1}{N} \sum_{n \in \mathcal{N}} P_{\text{U}}[n] \le \bar{P}_{\text{U}}, \tag{7.31e}$$

$$0 \le P_{\text{U}}[n] \le P_{\text{U}}^{\max}, \ \forall n \in \mathcal{N}, \tag{7.31f}$$

$$\frac{1}{N} \sum_{n \in \mathcal{N}} P_{\text{S}}[n] \le \bar{P}_{\text{S}}, \tag{7.31g}$$

$$0 \le P_{\text{S}}[n] \le P_{\text{S}}^{\max}, \ \forall n \in \mathcal{N}, \tag{7.31h}$$

$$R_{\text{S}}^{\text{LB}}[n] \ge r_{\text{S}}[n], \ \forall n \in \mathcal{N}, \tag{7.31i}$$

$$R_{\text{E}}^{\text{UB}}[n] \le r_{\text{E}}[n], \ \forall n \in \mathcal{N}, \tag{7.31j}$$

$$\frac{1}{N} \sum_{n \in \mathcal{N}} \left(\rho_0 d_{\text{SP}}^{-\varphi} P_{\text{S}}[n] + P_{\text{U}}[n] g_{\text{UP}}[n] \right) \le \varepsilon. \tag{7.31k}$$

It is easy to see that the objective function in (7.31) is a linear function. Therefore, we now concentrate on addressing the non-convex constraint (7.31i)–(7.31k).

Convexity of (7.31i): We introduce slack variables $z_{\text{S}}[n]$ and $t_{\text{S}}[n]$ such that

$$(7.31i) \Leftrightarrow \begin{cases} R_{\text{S}}^{\text{LB}}[n] \ge \log_2(1 + t_{\text{S}}[n]) \ge r_{\text{S}}[n], & (7.32a) \\[2mm] \dfrac{e^{-k} \gamma_0 d_{\text{SS}}^{-\varphi} P_{\text{S}}[n]}{\gamma_0 z_{\text{S}}^{-1}[n] P_{\text{U}}[n] + 1} \ge t_{\text{S}}[n], & (7.32b) \\[2mm] \left(x_{\text{S}} - x[n] \right)^2 + \left(y_{\text{S}} - y[n] \right)^2 + H^2 \ge z_{\text{S}}[n]. & (7.32c) \end{cases}$$

We first transform (7.32a) into a linear constraint by using the first-order Taylor expansion of $\log_2 \left(1 + t_{\text{S}}[n] \right)$ to find its lower bound around the point $t_{\text{S}}^{(i)}[n]$ at iteration i as in [30, Eq. (66)]:

$$R_{\text{S}}^{(i)}[n] \triangleq a \left(t_{\text{S}}^{(i)}[n] \right) - b \left(t_{\text{S}}^{(i)}[n] \right) \frac{1}{t_{\text{S}}[n]} \ge r_{\text{S}}[n], \ n \in \mathcal{N}, \tag{7.33}$$

where $a(t_S^{(i)}[n]) \triangleq \log_2(1 + t_S^{(i)}[n]) + \log_2(e)\frac{t_S^{(i)}[n]}{t_S^{(i)}[n]+1}$ and $b(t_S^{(i)}[n]) \triangleq \log_2(e)$
$\frac{(t_S^{(i)}[n])^2}{t_S^{(i)}[n]+1}$. Next, we rewrite (7.32b) as

$$t_S[n] \left(\gamma_0 P_U[n] + z_S[n] \right) \le e^{-k}\gamma_0 d_{SS}^{-\varphi} P_S[n] z_S[n] \tag{7.34}$$

and then exploit the inequality

$$xy \le 0.5 \left(\frac{y^{(i)}}{x^{(i)}}x^2 + \frac{x^{(i)}}{y^{(i)}}y^2 \right), \text{ for } x,y \in \mathbb{R}_+, x^{(i)}, x^{(i)} > 0$$

to convexify (7.34) as

$$\frac{1}{2}\frac{t_S^{(i)}[n]}{\gamma_0 P_U^{(i)}[n] + z_S^{(i)}[n]}\left(\gamma_0 P_U[n] + z_S[n] \right)^2$$
$$+ \frac{1}{2}\frac{\gamma_0 P_U^{(i)}[n] + z_S^{(i)}[n]}{t_S^{(i)}[n]}t_S^2[n] + \frac{e^{-k}\gamma_0 d_{SS}^{-\varphi}}{4}\left(P_S[n] - z_S[n] \right)^2$$
$$\le \frac{e^{-k}\gamma_0 d_{SS}^{-\varphi}}{4}\left(\left(P_S[n] + z_S[n] \right)^2 \right), n \in \mathcal{N}. \tag{7.35}$$

For constraint (7.32c), we also apply the first-order Taylor expansion for its left-hand side (LHS) to iteratively replace it by the following linear constraint:

$$\left(x_S - x^{(i)}[n] \right)^2 + 2 \left(x^{(i)}[n] - x_S \right) \left(x[n] - x^{(i)}[n] \right)$$
$$+ \left(y_S - y^{(i)}[n] \right)^2 + 2 \left(y^{(i)}[n] - y_S \right) \left(y[n] - y^{(i)}[n] \right) + H^2$$
$$\triangleq f_S^{(i)}(x[n], y[n]) \ge z_S[n], n \in \mathcal{N} \tag{7.36}$$

where $f_S^{(i)}(x[n], y[n])$ is the first-order approximation of the LHS of (7.32c) around the point $(x^{(i)}[n], y^{(i)}[n])$. We can easily see that (7.33), (7.35), and (7.36) are quadratic and linear constraints. Thus, (7.31c) is transformed into convex constraints.

Convexity of (7.31j): By introducing new variables $z_E[n], t_E[n]$ and $v[n]$, (7.31j) can be rewritten as follows:

$$(7.31j) \Leftrightarrow \begin{cases} R_E^{UB}[n] \le \log_2(1 + t_E[n]) \le r_E[n], & (7.37a) \\[2mm] \dfrac{\gamma_0 d_{SE}^{-\varphi} P_S[n]}{v[n]+1} \le t_E[n], & (7.37b) \\[2mm] v[n] \le \dfrac{\gamma_0 P_U[n]}{z_E[n]}, & (7.37c) \\[2mm] \left(x_E - x[n] \right)^2 + \left(y_E - y[n] \right)^2 + H^2 \le z_E[n]. & (7.37d) \end{cases}$$

It can be observed that all constraints in (7.37) except for (7.37d) are non-convex. Because the LHS of (7.37a) is a concave function, we first internally approximate it into a convex constraint at iteration i as

$$R_E^{(i)}[n] \triangleq \log_2(1 + t_E^{(i)}[n]) + \frac{\log_2(e)(t_E[n] - t_E^{(i)}[n])}{1 + t_E^{(i)}[n]} \leq r_E[n], \; n \in \mathcal{N}. \tag{7.38}$$

Similarly to (7.35), at the point $(P_S^{(i)}[n], v^{(i)}[n])$, constraint (7.37b) follows that

$$\frac{\gamma_0 d_{SE}^{-\varphi}}{2} \left(\frac{P_S^2[n]}{P_S^{(i)}[n](v^{(i)}[n] + 1)} + \frac{P_S^{(i)}[n](v^{(i)}[n] + 1)}{(v[n] + 1)^2} \right) \leq t_E[n]$$

and can be further approximated as

$$\frac{\gamma_0 d_{SE}^{-\varphi}}{2} \left(\frac{P_S^2[n]}{p_S^{(i)}[n](v^{(i)}[n] + 1)} + \frac{P_S^{(i)}[n]}{2v[n] - v^{(i)}[n] + 1} \right) \leq t_E[n], \; n \in \mathcal{N} \tag{7.39}$$

which is a convex constraint, where $(v[n] + 1)^2$ is lower bounded by $(v^{(i)}[n] + 1)(2v[n] - v^{(i)}[n] + 1)$ on the domain $2v[n] - v^{(i)}[n] + 1 > 0$. Constraint (7.37c) is equivalent to $v[n]z_E[n] \leq \gamma_0 p_U[n]$ and in the same manner as (7.35), we have

$$\frac{1}{2} \left(\frac{z_E^{(i)}[n]}{v^{(i)}[n]} v^2[n] + \frac{v^{(i)}[n]}{z_E^{(i)}[n]} z_E^2[n] \right) \leq \gamma_0 p_U[n], \; n \in \mathcal{N}. \tag{7.40}$$

Convexity of (7.31k): We first reformulated (7.31k) as

$$(7.31k) \Leftrightarrow \begin{cases} \dfrac{1}{N} \displaystyle\sum_{n \in \mathcal{N}} \left(\rho_0 d_{SP}^{-\varphi} P_S[n] + \rho_0 \dfrac{P_U[n]}{z_p[n]} \right) \leq \epsilon, & (7.41a) \\[4mm] \left(x_p - x[n] \right)^2 + \left(y_p - y[n] \right)^2 + H^2 \geq z_p[n], & (7.41b) \end{cases}$$

where $z_p[n]$ is new variable. By following the same step as in (7.39), constraint (7.41a) is convexified as

$$\frac{1}{N} \sum_{n \in \mathcal{N}} \left(\rho_0 d_{SP}^{-\varphi} P_S[n] + \frac{\rho_0}{2} \left[\frac{P_U^2[n]}{P_U^{(i)}[n]z_p^{(i)}[n]} + \frac{P_U^{(i)}[n]}{2z_p[n] - z_p^{(i)}[n]} \right] \right) \leq \epsilon. \tag{7.42}$$

Finally, constraint (7.41b) is internally approximated as

$$\begin{aligned}
& \left(x_p - x^{(i)}[n] \right)^2 + 2 \left(x^{(i)}[n] - x_p \right) \left(x[n] - x^{(i)}[n] \right) \\
& + \left(y_p - y^{(i)}[n] \right)^2 + 2 \left(y^{(i)}[n] - y_p \right) \left(y[n] - y^{(i)}[n] \right) \\
& + \left(h_p - h^{(i)}[n] \right)^2 + 2 \left(h^{(i)}[n] - h_p \right) \left(h[n] - h^{(i)}[n] \right) \\
& \triangleq f_p^{(i)}(x[n], y[n]) \geq z_p[n], \; n \in \mathcal{N}.
\end{aligned} \tag{7.43}$$

By incorporating all the above developments, we solve the following approximate convex program at iteration $(i + 1)$:

$$P4: \quad \max_{\mathbf{q}, \mathbf{p}, \mathbf{r}} \quad R_{\text{sec}}^{\text{LB}} \triangleq \frac{1}{N} \sum_{n \in \mathcal{N}} \left(r_{\text{S}}[n] - r_{\text{E}}[n] \right) \tag{7.44a}$$

$$\text{s.t.} \quad (x[1] - x_0)^2 + (y[1] - y_0)^2 \le L_{\max}^2, \tag{7.44b}$$

$$(x[n+1] - x[n])^2 + (y[n+1] - y[n])^2 \le L_{\max}^2, \ \forall n \in \mathcal{N}, \tag{7.44c}$$

$$(x_f - x[N])^2 + (y_f - y[N])^2 = 0, \tag{7.44d}$$

$$\frac{1}{N} \sum_{n \in \mathcal{N}} P_{\text{U}}[n] \le \bar{P}_{\text{U}}, \tag{7.44e}$$

$$0 \le P_{\text{U}}[n] \le P_{\text{U}}^{\max}, \tag{7.44f}$$

$$\frac{1}{N} \sum_{n \in \mathcal{N}} P_{\text{S}}[n] \le \bar{P}_{\text{S}}, \tag{7.44g}$$

$$0 \le P_{\text{S}}[n] \le P_{\text{S}}^{\max}, \tag{7.44h}$$

$$(7.33), (7.35), (7.36), (7.37d), (7.38), (7.39), (7.40), (7.42), (7.43),$$
$$\tag{7.44i}$$

where $\mathbf{z} \triangleq \{z_{\text{S}}[n], z_{\text{P}}[n], z_{\text{E}}[n]\}_{n \in \mathcal{N}}$, $\mathbf{t} \triangleq \{t_{\text{S}}[n], t_{\text{E}}[n]\}_{n \in \mathcal{N}}$ and $\mathbf{v} \triangleq \{v[n]\}_{n \in \mathcal{N}}$. Let $\mathbf{\Psi}^{(i)} \triangleq \{x^{(i)}[n], y^{(i)}[n], h^{(i)}[n], P_{\text{S}}^{(i)}[n], P_{\text{U}}^{(i)}[n], r_{\text{S}}^{(i)}[n], r_{\text{E}}^{(i)}[n], z_{\text{S}}^{(i)}[n], z_{\text{P}}^{(i)}[n],$ $z_{\text{E}}^{(i)}[n], t_{\text{S}}^{(i)}[n], t_{\text{E}}^{(i)}[n], v^{(i)}[n]\}, \ \forall n \in \mathcal{N}$, be the set of all the constant values that should be updated after each iteration. Initializing from a feasible point $\mathbf{\Psi}^{(0)}$, we solve (7.44) and update the involved variables until a stopping criterion is met. In particular, the proposed algorithm to solve (7.30) is summarized in Algorithm 7.1.

Convergence Analysis: It is true that the approximations of non-convex constraints $\{(7.31i), (7.31j), (7.31k)\}$ satisfy the properties of the IA method given in [29]. In other words, the proposed Algorithm 7.1 for solving (7.44) generates a sequence of non-decreasing objective values (i.e. $R_{\text{sec}}^{\text{LB},(i)} \ge R_{\text{sec}}^{\text{LB},(i-1)}$), which is provably monotonically convergent due to the constraints listed in (7.30). We can check that the solution obtained at each iteration (i.e. step 4 of Algorithm 7.1) meets the Karush–Kuhn–Tucker (KKT) conditions of (7.44). That is to say, the KKT conditions of (7.44) are also identical to those of (7.30), whenever $\mathbf{\Psi}^{(i)} = \mathbf{\Psi}^{(i-1)}$ [29, Theorem 1].

Algorithm 7.1: Proposed Iterative Algorithm to Solve (30)

1: **Initialization:** Set $i := 0$ and generate an initial feasible point $\Psi^{(0)}$ for all constraints in (44).
2: **repeat**
3:     Set $i := i + 1$;
4:     Solve (7.44) to find the optimal solution $\Psi^{(*)} \triangleq (\mathbf{q}^{(*)}, \mathbf{p}^{(*)}, \mathbf{r}^{(*)}, \mathbf{z}^{(*)}, \mathbf{t}^{(*)}, \mathbf{v}^{(*)})$;
5:     Update $\Psi^{(i)} := \Psi^{(*)}$;
6: **until** $\dfrac{R_{\mathrm{sec}}^{\mathrm{LB},(i)} - R_{\mathrm{sec}}^{\mathrm{LB},(i-1)}}{R_{\mathrm{sec}}^{\mathrm{LB},(i-1)}} \leq \epsilon_{\mathrm{tol}}$.

7.4 Numerical Results

In this section, we use computer simulations to evaluate system performance. The important parameters are provided in Table 7.1. The locations of ST, SU, and PR are set to $(0, 0, 0)$, $(300, 0, 0)$, and $(0, 250, 0)$, respectively. The UAV flies at a fixed altitude $H = 100$ m from the initial location at $(-100, 200, 100)$ to the final location at $(500, 200, 100)$. The location of Eve is assumed to be at $(150, 250, 0)$. Herein, we favor Eve by locating

Table 7.1 Simulation parameters

Parameter	Value
System bandwidth	10 MHz
Average interference power threshold at the PR, ε	-20 dBm
Noise power, σ^2	-70 dBm
Path loss exponent, φ	3
Number of time slots, N	500
Channel power gain at the reference distance, ρ_0	10 dB
Power budget at the ST, $P_{\mathrm{S}}^{\max}$	40 dBm
Power budget at the UAV, $P_{\mathrm{U}}^{\max}$	4 dBm
Average power limit at the ST, $\bar{P}_{\mathrm{S}}$	$P_{\mathrm{S}}^{\max}/2$
Average power limit at the UAV, $\bar{P}_{\mathrm{U}}$	$P_{\mathrm{U}}^{\max}/2$
Altitude of the UAV, H	100 m
Maximum speed of the UAV, $V_{\max}$	10 m/s
Error tolerance threshold, ϵ_{tol}	10^{-3}

it closer to the ST than the SR to demonstrate the effectiveness of the use of UAV-enabled JN. The other parameters are given in the captions of the figures.

For comparison, we consider the following four schemes:

- *Proposed method* (i.e. Algorithm 7.1).
- *Fixed power*: The solution of this scheme is also obtained by using Algorithm 7.1, where the transmit powers of the ST and UAV in each time slot are fixed as $\bar{P}_S$ and $\bar{P}_U$, respectively [26–28].
- *Straight line trajectory*: Under Algorithm 7.1, the UAV's trajectory is set as flying straight from the initial location to the final location [25–27].
- *No JN*: Under Algorithm 7.1, $P_U[n]$, $\forall n$ are set to 0 (i.e. without using UAV) [26, 27].

Figure 7.2 depicts the average secrecy rate of different methods versus flight time period, $T \in [0, 500 \text{ s}]$. We first observe that the "No JN" scheme is unable to provide a positive average secrecy rate since the ST–Eve link has a better channel quality compared to that of the ST–SR link. Such a result further confirms the importance of using UAV-enabled JN. As expected, the secrecy rates of other schemes increase as T increases. It is clear that the optimization of the UAV's trajectory is of crucial importance for the UAV to find an optimal location to interfere with the Eve, as evident from

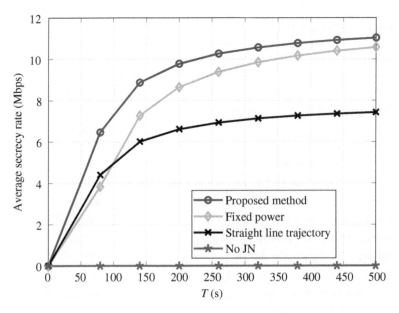

Figure 7.2 Average secrecy rate versus flight time period.

the average secrecy rates of the "Proposed method" and "Fixed power" when compared to "straight line trajectory." Overall, the proposed method of joint UAV's trajectory and power control obtains the best secrecy rate in all ranges of T. To summarize, the drawback of "straight line trajectory" scheme is not jointly optimizing the trajectory and thus the UAV not being able to fly to the best location to transmit JN to the Eve, while the limitation of "fixed power" scheme is that ST and UAV cannot adjust their transmission powers to optimal solutions.

In Figure 7.3, we illustrate the trajectories of UAV with different methods. Except for the straight line trajectory scheme, the trajectories are quite similar since the UAV tends to move closer to the Eve (but farther away from the SR to avoid a strong interference) and hover stationary above it to transmit the JN, whenever the average interference power at the PR is satisfied. Furthermore, when the time period T increases, the UAV has more time to stay on the optimal locations for jamming, which leads to a better secrecy rate, as shown in Figure 7.2.

In Figure 7.4, we plot the distance between the UAV and Eve as a function of T. Intuitively, the optimal distance of UAV–Eve is 100 m but not directly above the Eve due to a trade-off between degrading the wiretap channel

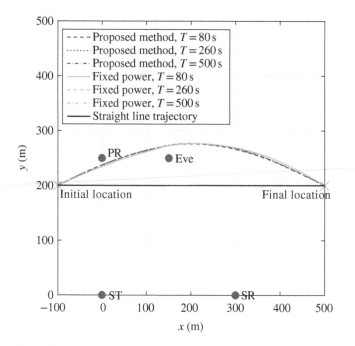

Figure 7.3 Trajectories of UAV with different methods.

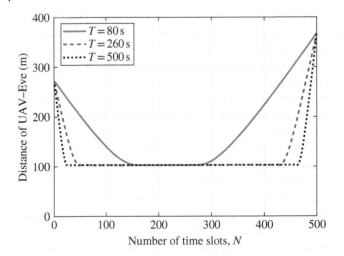

Figure 7.4 Distance between UAV and Eve.

and causing undesired interference to the SR and PR, at which the average secrecy rate is maximized. Another interesting observation is that the UAV remains stationary at the optimal location as long as possible, and this duration increases for a larger T to achieve a higher average secrecy rate.

Figure 7.5a shows the convergence behavior of Algorithm 7.1 for $T = 500\,$s, where the error tolerance for convergence is set as $\epsilon_{tol} = 10^{-3}$. It can be observed that the proposed algorithm monotonically improves the average secrecy rate at every iteration. In addition, Algorithm 7.1 requires only about eight iterations to achieve the optimal secrecy rate, which is also typical for other settings.

Finally, we show the influence of the average interference power thresholds at the PR on the secrecy rate of the secondary system with respect to the number of iterations in Figure 7.5b. It is observed that decreasing the average interference power threshold ϵ reduces the average secrecy rate. The performance gains achieved for a higher average interference power threshold are due to the fact that in this case, the UAV can fly to an optimal location more easily, and more transmit power at UAV and ST can be used.

7.5 Conclusion

In this chapter, we have studied the problem of maximizing the average secrecy rate of the CRN, where a UAV-enabled JN is deployed as a friendly jammer to degrade the decoding capability of Eve. The considered problem

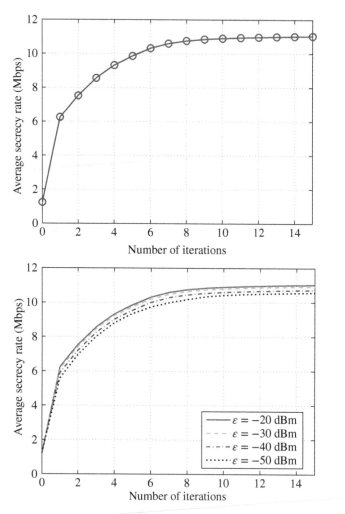

Figure 7.5 Convergence behavior. (a) Convergence behavior at $\varepsilon = -20\,\text{dBm}$. (b) Average secrecy rate versus threshold.

under the average interference power constraint at PR and transmit power constraints at the transmitters (ST and UAV) is non-convex. To solve the problem, we first transform it into a more tractable form and a safe design and then propose an inner approximation algorithm. In particular, we have shown that the problem can be approximated as a sequence of convex programs, which is efficiently solved using the existing solvers. The proposed algorithm outperforms all existing solutions in terms of the average secrecy rate and requires only a few iterations for convergence. Numerical results have revealed that jointly optimizing the transmit power at the transmitters and the UAV's trajectory can significantly enhance the average secrecy rate of the secondary system when compared to other designs.

Bibliography

1 Ericsson (2020). Mobile subscriptions outlook, June 2020.

2 Tran, D.H., Nguyen, V.D., Sumit, G. et al. (2020). UAV relay-assisted emergency communications in IoT networks: resource allocation and trajectory optimization. *arXiv:2008.00218.*

3 Federal Communication Commission (FCC) 02-155 (2002). Spectrum policy task force report, November 2002.

4 Nguyen, P.X., Pham, T.H., Hoang, T., and Shin, O. (2018). An efficient spectral leakage filtering for IEEE 802.11af in TV white space. *Proceedings International Conference on Recent Advances in Signal Processing, Telecommunications & Computing (SigTelCom)*, pp. 219–223.

5 Haykin, S. (2005). Cognitive radio: brain-empowered wireless communications. *IEEE Journal on Selected Areas in Communications* 23 (2): 201–220.

6 Dinh Tran, H., Trung Tran, D., and Choi, S.G. (2018). Secrecy performance of a generalized partial relay selection protocol in underlay cognitive networks. *International Journal of Communication Systems* 31 (17): e3806.

7 Nguyen, P.X., Nguyen, H.V., Nguyen, V., and Shin, O. (2019). UAV-enabled jamming noise for achieving secure communications in cognitive radio networks. *Proceedings of IEEE Consumer Communications and Networking Conference (CCNC)*, January 2019, pp. 1–6.

8 Hieu, T.D., Duy, T.T., and Choi, S.G. (2018). Performance enhancement for harvest-to-transmit cognitive multi-hop networks with best path selection method under presence of eavesdropper. *2018 20th International Conference on Advanced Communication Technology (ICACT)*, IEEE, pp. 323–328.

9 Tragos, E.Z., Zeadally, S., Fragkiadakis, A.G., and Siris, V.A. (2013). Spectrum assignment in cognitive radio networks: a comprehensive survey. *IEEE Communication Surveys and Tutorials* 15 (3): 1108–1135.

10 Shu, Z., Qian, Y., and Ci, S. (2013). On physical layer security for cognitive radio networks. *IEEE Network* 27 (3): 28–33.

11 Nguyen, V.-D., Duong, T.Q., Dobre, O.A., and Shin, O.-S. (2016). Joint information and jamming beamforming for secrecy rate maximization

in cognitive radio networks. *IEEE Transactions on Information Forensics and Security* 11 (11): 2609–2623.

12 Nguyen, V.-D., Duong, T.Q., Shin, O.-S. et al. (2017). Enhancing PHY security of cooperative cognitive radio multicast communications. *IEEE Transactions on Cognitive Communications and Networking* 3 (4): 599–613.

13 Zou, Y., Wang, X., and Shen, W. (2013). Physical-layer security with multiuser scheduling in cognitive radio networks. *IEEE Transactions on Communications* 61 (12): 5103–5113.

14 Zou, Y., Zhu, J., Yang, L. et al. (2015). Securing physical-layer communications for cognitive radio networks. *IEEE Communications Magazine* 53 (9): 48–54.

15 Wyner, A.D. (1975). The wire-tap channel. *The Bell System Technical Journal* 54 (8): 1355–1387.

16 Wu, Y., Schober, R., Ng, D.W.K. et al. (2016). Secure massive MIMO transmission with an active eavesdropper. *IEEE Transactions on Information Theory* 62 (7): 3880–3900.

17 Nguyen, V.-D., Nguyen, H.V., Dobre, O.A., and Shin, O.-S. (2018). A new design paradigm for secure full-duplex multiuser systems. *IEEE Journal on Selected Areas in Communications* 36 (7) 1480–1498.

18 Bassily, R., Ekrem, E., He, X. et al. (2013). Cooperative security at the physical layer: a summary of recent advances. *IEEE Signal Processing Magazine* 30 (5): 16–28.

19 Zeng, Y., Zhang, R., and Lim, T.J. (2016). Wireless communications with unmanned aerial vehicles: opportunities and challenges. *IEEE Communications Magazine* 54 (5): 36–42.

20 Tran, D.H., Vu, T.X., Chatzinotas, S. et al. (2020). Coarse trajectory design for energy minimization in UAV-enabled. *IEEE Transactions on Vehicular Technology* 69 (9): 9483–9496.

21 Zeng, Y., Zhang, R., and Lim, T.J. (2016). Throughput maximization for UAV-enabled mobile relaying systems. *IEEE Transactions on Communications* 64 (12): 4983–4996.

22 Sotheara, S., Aso, K., Aomi, N., and Shimamoto, S. (2014). Effective data gathering and energy efficient communication protocol in wireless sensor networks employing UAV. *2014 IEEE Wireless Communications and Networking Conference (WCNC)*, April 2014, pp. 2342–2347.

23 Tran-Dinh, H., Vu, T.X., Chatzinotas, S., and Ottersten, B. (2019). Energy-efficient trajectory design for UAV-enabled wireless communications with latency constraints. *2019 53rd Asilomar Conference on Signals, Systems, and Computers*, pp. 347–352.

24 Xie, P., Zhang, M., Zhang, G. et al. (2018). On physical-layer security for primary system in underlay cognitive radio networks. *IET Networks* 7 (2): 68–73.

25 Wang, Q., Chen, Z., Mei, W., and Fang, J. (2017). Improving physical layer security using UAV-enabled mobile relaying. *IEEE Wireless Communications Letters* 6 (3): 310–313.

26 Li, A., Wu, Q., and Zhang, R. (2018). UAV-enabled cooperative jamming for improving secrecy of ground wiretap channel. *IEEE Wireless Communications Letters* 8 (1) 181–184.

27 Zhang, G., Wu, Q., Cui, M., and Zhang, R. (2017). Securing UAV communications via trajectory optimization. *Proceedings of IEEE Global Communications (IEEE GLOBECOM)*, Singapore, December 2017, pp. 1–6.

28 Lee, H., Eom, S., Park, J., and Lee, I. (2018). UAV-aided secure communications with cooperative jamming. *IEEE Transactions on Vehicular Technology* 67 (10)9385–9392.

29 Marks, B.R. and Wright, G.P. (1978). A general inner approximation algorithm for nonconvex mathematical programs. *Operations Research* 26 (4): 681–683.

30 Nguyen, V.-D., Tuan, H.D., Duong, T.Q. et al. (2017). Precoder design for signal superposition in MIMO-NOMA multicell networks. *IEEE Journal on Selected Areas in Communications* 35 (12): 2681–2695.

8

IRS-Assisted Localization for Airborne Mobile Networks

Olaoluwa Popoola[1], Shuja Ansari[1], Rafay I. Ansari[2], Lina Mohjazi[1], Syed A. Hassan[3], Nauman Aslam[2], Qammer H. Abbasi[1], and Muhammad A. Imran[1]

[1] *James Watt School of Engineering, University of Glasgow, Glasgow, UK*
[2] *Department of Computer and Information Science, Northumbria University, Newcastle upon Tyne, UK*
[3] *School of Electrical Engineering and Computer Science (SEECS), National University of Sciences and Technology (NUST), Islamabad, Pakistan*

8.1 Introduction

The increase in the number of connected devices and the vision of 5G to provide ultra-reliable low-latency communication (URLLC) has introduced several opportunities with regard to applications that impact health, agriculture, and transport, among others. The key enabling technologies of 5G such as Internet-of-Things (IoT) envision a massive connectivity of devices along with ensuring key performance metrics such as low delay, high data rates, and enhanced reliability. The limitation of spectrum resources has given impetus to exploring new spectrum opportunities lying within the 30–300 GHz millimeter wave (mmWave) band. New technologies such as unmanned aerial vehicles (UAVs)-based base stations (BS) have been proposed to assist the traditional cellular network, especially for disaster area scenarios.

UAV-BS possess the capability to move in three-dimensional space, making them a suitable candidate for on-demand deployment. In a scenario where the cellular network is either choked due to high demand or is undergoing any connectivity issues, the UAV-BS can be deployed to provide on-demand services. The deployment of UAV-BS can help in sharing the burden of the cellular network and provide services to the users. However, there are several constraints associated with the UAV operation, with the most critical being the battery time. The UAV-BS has to replenish its energy resources after a certain period of operation, which can compromise the sustainability of the transmissions. Moreover, the selection of the frequency band employed by the UAV-BS is important for avoiding any interference issues. Viewing the challenges associated with utilizing the UAV-BS, there is considerable interest in the research community to analyze the UAV-BS operation from different perspectives and to ensure that it meets the performance criteria of 5G and beyond (B5G).

Recently, intelligent reflecting surfaces (IRSs) have been explored with a vision of providing smart wireless connectivity. The growth in the number of devices and the strict requirements for latency and rate have highlighted the need for improving the link quality in different network environments. The BSs have utilized technologies such as multiple-input multiple-output (MIMO) to optimize the transmissions and provide transmission diversity. However, the propagation environment with different factors such as reflection, diffraction, and scattering is random in nature and it cannot be managed effectively through the BS operation optimization. IRSs provide the opportunity to assist the transmission as they are reconfigurable and can provide performance improvement by different functionalities such as reflection, absorption, and phase change. IRS is a sheet with multiple scattering elements, where the intelligence of the surface comes from phase control of all the scattering elements to realize a desired operation. The manipulation of electromagnetic (EM) waves can be conducted to avoid interference by managing the phase of each wave to interact destructively, or to enhance the signal strength by having a constructive interaction of multi-paths [1].

In this chapter, we explore a network framework that involves UAVs and IRSs for supporting localization. The localization of devices is a part of the network contextual information that is required to ensure seamless connectivity. Localization has been explored for wireless sensor networks to ensure reliable hop-to-hop connectivity. Several approaches have been proposed to improve the localization accuracy. Location information is important for network planning and dynamic resource allocation for improving link quality. The localization process comprises of collecting the location information and estimating the location based on different parameters [2]. Localization gains even more importance in network environments involving mobility such as vehicle-to-vehicle (V2V) networks. The dynamic nature of such networks makes the localization information precious with regard to maintaining a strong transmitter–receiver link.

8.1.1 Related Work

8.1.2 Unmanned Aerial Vehicles

UAVs have gained significant interest of the research community due to their mobility and ease of deployment. Users can be assisted in areas where the cellular network is inaccessible or is unable to cater to the rising demands. The UAV-BSs can be deployed to share the burden of connectivity with the cellular network. UAV-BSs gain particular importance in scenarios where direct line-of-sight (LoS) links are difficult to establish due to physical structures. The UAV-BSs can adjust their location to provide the best connectivity and assist any users that are facing connectivity issues. The UAV-BSs can improve the network coverage by avoiding any coverage holes. The UAVs can also act as a relay for any terrestrial stations [3].

Energy-efficient (EE) communication is one of the key metrics associated with the UAV-BSs due to the limitations of battery resources. Several EE optimization algorithms have been proposed for UAV-BSs to prolong the flight time. The transmit power of UAV-BSs can be optimized and the scheduling can be performed to ensure EE. The trajectory of the UAV-BS directly impacts the network performance in terms of reliability and delay. A joint optimization of the aforementioned factors can ensure seamless connectivity and help in realizing the performance metrics for B5G. Multiple UAV-BSs can be deployed to provide connectivity in a region, depending on the service requirements [4].

Three-dimensional movement and availability of miniaturized transmitter/receiver antennas make the UAV-BS realizable for both uplink and downlink transmissions [5]. Trajectory optimization of the UAV-BS can impact the network throughput and the energy utilization. It is important to take the interference performance into consideration, especially for scenarios where multiple transmitters are present. Interference mitigation techniques have been proposed to improve the transmission reliability. Owing to the presence of a dominant LoS link, the UAV-BSs may suffer from higher interfering signals from the terrestrial BSs. The interference mitigation techniques used for traditional cellular networks may not be suitable for UAV-based networks [6]. The deployment planning of UAVs can be conducted based on the density of devices on the ground that require connectivity. In a nutshell, UAV-BSs can be of significant assistance in supporting cellular networks and realizing different applications [7].

8.1.3 Intelligent Reflecting Surface

Intelligent (reconfigurable) surfaces can assist point-to-point communications, especially in scenarios where direct LoS connections are not possible due to physical hurdles. The use of IRS gains further importance in case of higher frequency bands such as mmWaves, where highly directive links are required and any physical obstruction can hinder the transmissions. The IRS can help in establishing a non-LoS link and provide alternate paths for the transmission. IRS is a planar array of reflecting elements, where the elements are reconfigurable. The reconfiguration can be conducted through a controller, which changes the surface impedance of the IRS and can alter the phase of the reflected electromagnetic wave [8].

The elements in the IRS can introduce phase shifts that are helpful in overcoming the limitations of propagation environment, i.e. exploiting the multipaths to enhance the received signal strength (RSS) [9]. Other advantages of utilizing IRSs include an increase in the overall coverage area and signal strength. IRSs can also increase the energy efficiency as a lower transmit power is required [8]. The strip-shaped design of the IRSs makes them a suitable candidate for deployment in both indoor and outdoor networks. Several works have appeared in the literature focusing on optimizing the parameters of the IRS to optimize the RSS and reduce the overall energy consumption [10, 11]. Latency minimization by utilizing the IRSs has also been a key research problem, especially with regard to meeting the URLLC requirement for 5G networks. The absorption function of the IRS is particularly helpful for introducing the physical layer security [12], by avoiding the message delivery to eavesdropping nodes [13].

The operation of the IRS seems synonymous to a relay but IRSs possess different capabilities as compared to a relay. The IRS does not amplify the signals as the traditional arrays can amplify, but it can enhance the signal strength at the receiver by exploiting the array power gain. A comparison of the traditional relays and the IRSs highlights the performance gains that can be achieved in terms of signal-to-noise ratio (SNR) [14]. One of the challenges of utilizing IRSs is to gather the channel state information (CSI), as the CSI is required to benefit from the beamforming gain provided by the IRS. However, the CSI overhead can be taxing in terms of processing and there is a need to minimize the overhead [15].

8.2 Intelligent Reflecting Surfaces in Airborne Networks

Considering the opportunities put forward by the revolutionary IRS technology, its deployment in airborne platforms is attracting growing interest from academia and industry. This is driven by its potential in offering enhanced coverage extension and communication reliability, as well as a low-cost approach for network densification. To this effect, IRS deployment to support future airborne networks is envisioned to be realized by two main streams. The first is the integration of IRS with aerial platforms, which can be done in a number of methods based on the platform's shape. In the second, the aerial network may be supported by IRS instrumented on building facades and can be programmed remotely or configured by BS in real-time [16]. In the following subsections, Sections 2.1 and 2.2, the two methods of IRS deployments in airborne networks, distinguished as aerial networks with integrated IRS and IRS-assisted aerial networks, are presented.

8.2.1 Aerial Networks with Integrated IRS

Owing to the advanced features of aerial platforms and the numerous afore-mentioned IRS capabilities, it is envisioned that aerial platforms will be integrated with IRS. This can be achieved by installing the IRS as a separate horizontal surface at the bottom of the platform [17]. The availability of CSI and the accurate configuration of IRS are crucial for performing the communication functions properly. In such a scenario, two levels of control management can exist, namely, a ground control station and an onboard aerial platform control.

The fundamental element in the controller of the ground control station is the processing unit whose role is to analyze the data, which is sensed from the aerial platform, and the access control and users' localization conditions, which are acquired during the information exchange with the ground BSs and/or gateways. The processing unit is also responsible for the joint management of the flying and communication functions of the aerial platform. Additionally, it estimates the CSI and angle of arrival (AoA) between users and the aerial platform to provide the best IRS configuration parameters [17].

The onboard aerial platform controller is expected to comprise of two units, namely, the flight control and the IRS control units. The former maintains the stabilization of the platform based on commands received from the ground station. The latter translates the received optimized IRS configuration into a switch control activation map, which is then applied to the IRS for manipulating the incident signals. In what follows, three potential use cases for integrating IRS in future airborne platforms are discussed.

8.2.1.1 Integration of IRS in High-Altitude Platform Stations (HAPSs) for Remote Areas Support

Recently, high-altitude platform stations (HAPSs) were introduced as viable aerial network components to connect remote users in future communication networks using cost-effective technologies and materials [18]. HAPSs are quasi-stationary nodes that operate in the stratosphere at a fixed point relative to the earth. Since HAPS consists of large surfaces suitable to accommodate solar panel films, the primary means of providing energy for HAPS is solar power coupled with energy storage. Energy consumption in HAPS is due to propulsion, stabilization, and communication operations. In certain application scenarios, other power sources are included to augment the harvested solar energy, which may not be sufficient for specific operations. Nonetheless, energy efficiency in HAPS is still considered as an open research problem, especially since it directly impacts the flight duration. Apart from energy consumption, reliable communication and light payload should be considered in designing HAPS deployments.

In this context, the integration of IRS in HAPS is introduced for wireless traffic backhauling from remote area BSs. In such a scenario, the traffic of a cluster of users is handled by BSs and transmitted to the HAPS. Then through the IRS equipped on HAPS, the received signals are intelligently reflected in the direction of a gateway station, which is part of the core network. If the gateway station does not fall within the HAPS coverage range, the IRS can be reconfigured to reflect the signals toward a nearby HAPS, which in its turn assists in the signal transmission in a multi-hop fashion. The benefits of integrating IRS in HAPS are threefold; (i) reduced power consumption required for communication functionalities, due to the signals being reflected in a nearly passive fashion, (ii) lighter payload since the IRS is made of thin lightweight materials, (iii) minimized energy required for stabilization as a result of reduced communication components and payload. Additionally, HAPSs integrated with IRS are significantly promising in achieving enhanced energy efficiency compared to conventional HAPS, due to the low-power switches embedded in the IRS [19].

8.2.1.2 Integration of IRS in UAVs for Terrestrial Networks Support

The benefits of IRS in terms of agility, flexibility, and rapid deployment can be reaped in UAV-assisted terrestrial cellular networks in a cost-effective manner. More specifically, by mounting IRS on a swarm of UAVs, an intermediate reflection layer can be formed between ground BSs and isolated users. This not only enables a smooth mechanical movement of the formed IRS layer but also provides a digital means for tuning the signals incident on the IRS. This enables the coverage holes in the network to be reduced and the operators' revenue per user to be boosted.

The integration of IRS with UAVs is also envisioned to provide a potential solution for the signal blockage problem that would be experienced in future communication networks operating in higher frequencies, such as mmWave and free space optics (FSO) [20]. In this case, UAVs equipped with IRS can deliver a 360° signal reflection approach and can potentially reduce the number of reflections compared to terrestrial IRS. Furthermore, UAVs mounted with IRS can participate in balancing the traffic of congested cells by enabling the delivery of new signals to users served by highly loaded ground BSs.

On another front, the reliability of signal reception may be significantly improved by the swarm of UAVs integrated with IRS. This stems from the fact that UAVs with IRS provide an efficient mechanism to combat channel impairments by offering spatial diversity. In this setup, a user receives multiple copies of its intended signal over multiple transmission paths created by numerous reflections through the UAV swarm. Furthermore, the data throughput in the network can be notably improved since multiple independent streams of information can be transmitted by the IRS mounted on the UAV and then retrieved by the user. It is also worth noting that the IRS signal absorption feature can be leveraged to recharge the UAV.

8.2.1.3 Integration of IRS with Tethered Balloons for Terrestrial/Aerial Users Support

Tethered UAVs have been recently introduced to overcome the shortcomings of conventional UAVs [21]. These shortcomings include their suitability for short-term deployments due to their limited onboard energy, and the limited reliability and capacity of wireless backhauling. On the contrary, tethered UAVs enjoy longer flight duration and reliable backhaul links since the tether provides the UAV with both power and data. Nonetheless, the deployment and operational costs of tethered UAVs is higher compared to non-tethered UAVs since the energy consumption of the former is higher compared to that of the latter.

As a cost-effective alternative, the integration of IRS in tethered balloons is proposed as a wireless access point to enhance the connectivity and capacity of urban users [17]. It is envisioned in this scenario that IRS mounted on the balloon generates multiple beams with different directions to serve multiple users, thereby improving the spectral efficiency of the system by introducing many near-field strong LoS links to users. Moreover, unlike ground BSs employed with down-tilted antennas, the integration of IRS with tethered balloons enables reliable connectivity to UAV users and therefore, opens the door for enhanced network densifications.

8.2.2 IRS-Assisted Aerial Networks

Since IRS-assisted communications can drastically improve the coverage and capacity of existing wireless networks, there is a growing interest in investigating the deployment of IRS to assist aerial-to-ground communications. Under this setup, the IRS is configured with appropriate phase shifts to reflect the impinging BS signals toward a targeted UAV or vice versa.

A recent study demonstrates that IRSs are potential candidates for enhancing cellular communications for UAVs, which currently suffer from poor signal strength due to the down-tilt of BS antennas optimized to serve ground users [22]. In this framework, an IRS mounted on building facades can be configured remotely by cellular BS to coherently direct the reflected EM waves toward specific UAVs with the objective of increasing their RSS s. For a downlink (BS to UAV) scenario, it is shown that the received cellular signal strengths at the UAVs, flying above the BS antenna, is considerably enhanced by a small IRS patch. Furthermore, an optimal selection of the IRS deployment altitude and distance from BS may maximize the gain from IRS. This constitutes a promising mechanism to improve the cellular coverage and the flying UAVs throughput.

The assistance of IRS is also shown to be beneficial in substantially boosting the communication quality of UAV communication systems operating in complex urban environments. In such environments, the LoS link between the UAV and the ground user may be blocked, resulting in the channel quality being severely deteriorated. To this effect, a mobile UAV communicates with a ground user along its planned trajectory, and its transmitted signal is reflected to the user via the IRS [23]. It is demonstrated that the joint design of UAV trajectory and passive beamforming, which is subject to practical UAV mobility and the IRS's phase-shift constraints, results in a maximized average achievable rate.

Furthermore, IRS is introduced to assist UAVs that serve multiple ground users [24]. Particularly, passive beamforming supported by an IRS coated on the surface of a building can reflect the signals transmitted from a UAV. This allows for high flexibility to be achieved in minimizing the overall power consumption of the system by jointly optimizing the UAV trajectory and the resource allocation while meeting the individual minimum data rate requirement of each user and the limited energy budget of the IRS. Compared to conventional UAV networks where an IRS does not exist, this topology delivers an increased total power saving.

Finally, the deployment of IRS is proposed to improve the communication quality in aerial–ground multiuser systems [25]. This may be achieved by jointly optimizing the phase shifts at the IRS and the UAV trajectory to maximize the sum rate. In this scenario, the IRS deployed on the building facades improves the propagation environment by addressing the challenge of the LoS channels blockage in the aerial–ground communication.

8.3 Localization Using IRS

Use of IRS in multiple scenarios as discussed in the Section 8.2 opens up multiple research avenues in data transmission and effective user management. In mobile networks, the localization of users and mobile base stations is imperative for smooth and successful data transmission. In this regard, this section develops an overall system model for designing an IRS-enhanced localization system. The model will consist of the IRS for improving the localization accuracy, 5G small cell (SC), which is used as an omnidirectional wireless signal source, and a time of arrival (ToA) and RSS localization algorithm that provides the location of the UAV based on the strength of power received from the SC and the strength of reflected power received from the IRS.

In order to uniquely identify the source of an incoming signal at user equipment (UE), we propose to utilize pilot sub-carriers of an orthogonal frequency division multiplexing (OFDM) block in the downlink direction. The authors in [26, 27] proposed a similar concept of utilizing pilot sub-carriers in the uplink transmission for channel estimation and reflection optimization. We consider a similar technique for the downlink direction with a proposition of adding an IRS respective phase shift in a pilot sub-carrier. From Figure 8.1, consider an arbitrary reflecting unit $l \in \mathbf{M}$, whose reflection coefficient is given by $\phi_l = \beta_l e^{j\theta_l}$, if the reflecting unit shifts the phase by $\theta_l = \frac{2\pi}{L}$, where L is the total number of IRSs; on receiving this pilot symbol, the UE will be able to identify the reflecting surface for that signal.

Hence, with enhanced OFDM for IRS, a simple downlink pilot sub-carrier can be utilized for coarse localization in addition to channel estimation and reflection optimization. This localization information or similar principles

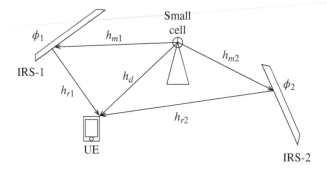

Figure 8.1 Localization using IRS model with two IRSs and one SC.

can be applied for identifying the UE's location. With successful identification of IRS, the UE can perform on board triangulation techniques for precise positioning. To simplify the evaluation of the system, we consider a space with single SC and multiple IRSs.

8.3.1 IRS Localization with Single Small Cell (SSC)

In an environment with single small cell, all IRSs receive information from the same SC and reflect the information toward the UAV. Figure 8.2 shows an illustration of an outdoor environment with an single small cell (SSC), four IRSs, and a UAV that is local to that environment. Since the positions of the SSC and IRSs are constant relative to the ground truth, we can assume that the positions of the SSC and IRSs are known and that there are N IRSs participating in localization.

8.3.1.1 IRS Localization Using RSS with an SSC

To localize the UAV using RSS, the power received at the UAV from the SSC is used to compute the distance between the UAV and the SSC and the power received from an individual IRS is used to compute the distance between the UAV and that IRS. Power received at the UAV that is used for localization could either come from the SSC or from the IRS. To compute the distance based on power received directly for the SSC, we consider the UAV at a distance d from the SSC, using the Hata Okumara log-distance model [28]; if the transmitted power level is P_{tx} (dBm), and the received power level at the

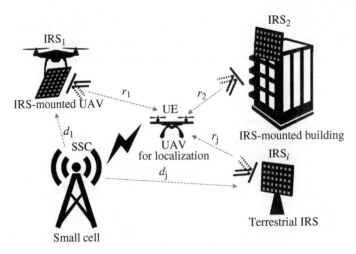

Figure 8.2 Localization of a UAV using multiple IRSs.

UAV is P_{rx} (dBm), then the distance d can be computed as

$$\log d = \frac{1}{10n}(P_{tx} - P_{rx} + G_{tx} + G_{rx} - X_\alpha + 20\log\lambda - 20\log(4\pi)), \quad (8.1)$$

where G_{tx} is the antenna gain of the transmitter, G_{rx} is the antenna gain of the receiver, X_α is the normal random variable with a standard deviation α, λ(m) is the wavelength of the signal and n is a measure of the influence (attenuation and scattering) of obstacles that include walls, doors, and partitions. Using (8.1), the distance between a UAV and a small cell can be computed. However, in order to compute distances between the IRSs and the UAV, we must obtain information of the power received at the IRS and the power that the IRS transmits toward the UAV after reflection and scattering.

The IRS has the advantage of providing a highly directed beam of reflected wave toward the receiver. In addition, as explained in Section 8.3, the signal reflected by each IRS is unique. Using RSS to compute the distance r from the IRS to the UAV, for a transmitted power P_t from the SC, by the property of IRSs, the reflected received power at a far field distance r by the jth IRS, which is at a distance d_j from the small cell $P_r(P_t, d_j, r, \theta_s)$, is given in [29] as

$$P_r(P_t, d_j, r, \theta_s) = \frac{1}{2\eta}S_{IRS}\left(r, \theta_s, \frac{P_tG_t\eta}{2\pi d_j^2}\right)\left(\frac{\lambda^2}{4\pi}G_r\right), \quad (8.2)$$

where θ_s is the angle of the scattered wave from the IRS, η is the characteristic impedance of the IRS medium, G_t, G_r are the transmitter and receiver antenna gains, λ is the wavelength of the electromagnetic wave, θ_r is the direction of the main beam of reflected signal, and $S_{IRS}(\cdot)$ is the squared magnitude of the scattered wave given as

$$S_{IRS}(r, \theta_s, E_i^2) = \left(\frac{ab}{\lambda}\right)^2 \frac{E_i^2\cos^2(\theta_i)}{r^2}\left(\frac{\sin(\frac{\pi b}{\lambda}(\sin(\theta_s) - \sin(\theta_r)))}{\frac{\pi b}{\lambda}(\sin(\theta_s) - \sin(\theta_r))}\right)$$

$$(8.3)$$

at a far-field distance $r \geq \frac{2\max(a^2, b^2)}{\lambda}$ where a and b are the dimensions of the IRS.

Using (8.2) and (8.3) when the main beam is in the desired direction toward the UAV such that $\theta_r = \theta_s$, we can compute the distance between the jth IRSs and the UAV as

$$r_j = \frac{ab\cos(\theta_i)}{4\pi d_j}\sqrt{\frac{P_tG_tG_r}{P_r}}, \quad \forall j \in 1, 2, \ldots, N. \quad (8.4)$$

Two important facts that can be extracted from (8.4) are that the distance between the jth IRS and that the UAV is proportional to the incident angle between the SSC and the IRS and inversely proportional to the received power at the UAV and also inversely proportional to the distance between the SSC and the IRS. Under ideal conditions, determining the location of the UAV from the set of equations (8.1) and (8.4), where the locations of the IRSs are known, is a trilateration problem and the solution is trivial. However in practice, the conditions are far from ideal, which leads to major challenges with the use of the RSS localization algorithm. These challenges are in the form of interference and fading due to multipath propagation.

8.4 Research Challenges

The development and commercialization of airborne networks and subsequent integration of IRS in aerial settings is still far from reality due to numerous open research challenges. In this section, we outline some of the main challenges that need to be overcome for the pursuit of IRS-based localization and their use in airborne wireless networks.

8.4.1 Challenges in UAV-Based Airborne Mobile Networks

First and foremost is the design challenge associated with getting UAV-based ANs up and running. Having a mobile base station comes with the question of energy consumption and backhaul links. In the case of tethered UAV-BS, we can overcome this challenge; however, for more robust and remote link provisioning, the onboard power constraints are the biggest obstacles in the realization of UAV-BS. In addition to power, the networking mechanisms are required to be impervious to the vibrant network dynamics of ANs. Additional and new network service protocols such as address and session management will be required for the fast changing topologies of ANs. Furthermore, network integration and configuration is required to be fast, robust, and more importantly highly autonomous, since the dynamics of UAVs and the subsequent stringent application requirements will come along a different set of data transmission challenges in terms of interference and path losses.

To overcome the challenges that come with fast changing environments, the robust autonomy of self-organizing networks will complement the UAV-BSs very well. The autonomous configuration of networks will not only address the connectivity and data transmission but will also require new security paradigms. Similarly, this autonomous nature will also play its role in selecting appropriate links and interfaces robustly for adding an additional layer of reliability.

8.4.2 Challenges in IRS-Based Localization

Since first-gen IRSs are passive antenna elements, there are multiple challenges in achieving the localization as proposed in this chapter. Utilization of pilot symbols for the identification of individual IRSs can be an obstacle in current implementations due to pilot contamination. The large number of reflective elements, resembling passive antennas, alongside massive distributed deployment in an area can pose complicated pilot contamination. Therefore, foremost research and comprehensive studies are required for pilot decontamination in IRS-assisted systems. For example, in [30] the authors outline that pilot decontamination can be reached via efficient pilot assignment schemes with the objective of maximizing the minimum average signal to interference ratio among served users. While arranging these pilots, localization aspects as proposed in this chapter should be kept in mind. This will also in turn improve channel prediction and enhance accuracy.

Another important aspect of IRS is the control signaling for IRS microcontroller that changes the reflective coefficient and ultimately the phase of incoming signals. Having these IRS mounted in HAPS or UAVs will pose a challenge of controlling the elements in real-time. Effective backhaul or fronthaul solutions will be required for an efficient control of IRS. Similarly, with a large number of IRSs present in an environment, the distributed operation will also pose a challenge for real-time interference mitigation, programmability, large-scale configuration, and processing and pre-coding operations. Since IRS operations are mainly just at the physical layer, security and privacy concerns are also required to be tackled.

8.5 Summary and Conclusion

To accommodate the expansion of future mobile networks, airborne base stations based on UAVs and IRSs are some of the ways researchers are putting their efforts to sustain the global connectivity. This chapter adds to the current knowledge on IRSs and ANs in line with localization aspects. Related works around IRS- and UAV-based Base Stations are elaborated, followed by detailing the underlying opportunities in these domains. Integration of IRS in AN is discussed and various scenarios are presented. An IRS-based localization model is proposed for ANs along with some mathematical modeling of the proposed system. Finally, some future research challenges presenting research opportunities are included in the last section.

Concepts and techniques discussed in this chapter are in early stages of evaluation and require more development for them to be realized in a practical setting. For future works, the proposed localization model will be evaluated using intensive simulations while pilot symbol based identification will be implemented on a hardware test bed utilizing multiple IRSs and an emulated base station.

Bibliography

1 Gong, S., Lu, X., Hoang, D.T. et al. (2020). Toward smart wireless communications via intelligent reflecting surfaces: a contemporary survey. *IEEE Communications Surveys Tutorials* 22 (4): 2283–2314.

2 Shi, X., Yu, D., and Zhang, W. (2020). Quantitative relationship between localization accuracy and location privacy level in wireless localization system. *IEEE Signal Processing Letters* 27: 1055–1059.

3 Ahmed, S., Chowdhury, M.Z., and Jang, Y.M. (2020). Energy-efficient UAV relaying communications to serve ground nodes. *IEEE Communications Letters* 24 (4): 849–852.

4 Zhang, S., Zhang, H., Di, B. et al. (2019). Cellular UAV-to-X communications: design and optimization for multi-UAV networks. *IEEE Transactions on Wireless Communications* 18 (2): 1346–1359.

5 Hua, M., Yang, L., Wu, Q. et al. (2020). 3D UAV trajectory and communication design for simultaneous uplink and downlink transmission. *IEEE Transactions on Communications* 68 (9): 5908–5923.

6 Mei, W. and Zhang, R. (2020). UAV-sensing-assisted cellular interference coordination: a cognitive radio approach. *IEEE Wireless Communications Letters* 9 (6): 799–803.

7 Lai, C., Chen, C., and Wang, L. (2019). On-demand density-aware UAV base station 3D placement for arbitrarily distributed users with guaranteed data rates. *IEEE Wireless Communications Letters* 8 (3): 913–916.

8 Ye, J., Guo, S., and Alouini, M.S. (2020). Joint reflecting and precoding designs for SER minimization in reconfigurable intelligent surfaces assisted MIMO systems. *IEEE Transactions on Wireless Communications* 19 (8): 5561–5574.

9 Yu, G., Chen, X., Zhong, C. et al. (2020). Design, analysis, and optimization of a large intelligent reflecting surface-aided B5G cellular internet of things. *IEEE Internet of Things Journal* 7 (9): 8902–8916.

10 Abdullah, Z., Chen, G., Lambotharan, S. et al. (2020). A hybrid relay and intelligent reflecting surface network and its ergodic performance analysis. *IEEE Wireless Communications Letters* 9 (10): 1653–1657.

11 Bai, T., Pan, C., Deng, Y. et al. (2020). Latency minimization for intelligent reflecting surface aided mobile edge computing. *IEEE Journal on Selected Areas in Communications* 38 (11): 2666–2682.

12 Wang, H.M., Bai, J., and Dong, L. (2020). Intelligent reflecting surfaces assisted secure transmission without Eavesdropper's CSI. *IEEE Signal Processing Letters* 27: 1300–1304.

13 Shen, H., Xu, W., Gong, S. et al. (2019). Secrecy rate maximization for intelligent reflecting surface assisted multi-antenna communications. *IEEE Communications Letters* 23 (9): 1488–1492.

14 Björnson, E. and Sanguinetti, L. (2020). Power scaling laws and near-field behaviors of massive MIMO and intelligent reflecting surfaces. *IEEE Open Journal of the Communications Society* 1: 1306–1324.

15 Wang, Z., Liu, L., and Cui, S. (2020). Channel estimation for intelligent reflecting surface assisted multiuser communications: framework, algorithms, and analysis. *IEEE Transactions on Wireless Communications* 19 (10): 6607–6620.

16 Yang, L., Meng, F., Zhang, J. et al. (2020). On the performance of RIS-assisted dual-hop UAV communication systems. *IEEE Transactions on Vehicular Technology* 69 (9): 10385–10390.

17 Alfattani, S., Jaafar, W., Hmamouche, Y. et al. (2020). Aerial platforms with reconfigurable smart surfaces for 5G and beyond. arXiv preprint arXiv:200609328.

18 Alzenad, M., Shakir, M.Z., Yanikomeroglu, H. et al. (2018). FSO-based vertical backhaul/fronthaul framework for 5G+ wireless networks. *IEEE Communications Magazine* 56 (1): 218–224.

19 Basar, E., Di Renzo, M., De Rosny, J. et al. (2019). Wireless communications through reconfigurable intelligent surfaces. *IEEE Access* 7: 116753–116773.

20 Zhang, Q., Saad, W., and Bennis, M. (2019). Reflections in the sky: millimeter wave communication with UAV-carried intelligent reflectors. *2019 IEEE Global Communications Conference (GLOBECOM)*, pp. 1–6.

21 Kishk, M.A., Bader, A., and Alouini, M.S. (2020). On the 3-D placement of airborne base stations using tethered UAVs. *IEEE Transactions on Communications* 68 (8): 5202–5215.

22 Ma, D., Ding, M., and Hassan, M. (2020). Enhancing cellular communications for UAVs via intelligent reflective surface. *2020 IEEE Wireless Communications and Networking Conference (WCNC)*, pp. 1–6.

23 Li, S., Duo, B., Yuan, X. et al. (2020). Reconfigurable intelligent surface assisted UAV communication: joint trajectory design and passive beamforming. *IEEE Wireless Communications Letters* 9 (5): 716–720.

24 Cai, Y., Wei, Z., Hu, S. et al. (2020). Resource allocation for power-efficient IRS-assisted UAV communications. *2020 IEEE International Conference on Communications Workshops (ICC Workshops)*, pp. 1–7.

25 Li, J. and Liu, J. (2020). Sum rate maximization via reconfigurable intelligent surface in UAV communication: phase shift and trajectory optimization. *2020 IEEE/CIC International Conference on Communications in China (ICCC)*, pp. 124–129.

26 Zheng, B. and Zhang, R. (2019). Intelligent reflecting surface-enhanced OFDM: channel estimation and reflection optimization. *IEEE Wireless Communications Letters* 9 (4): 518–522.

27 Yang, Y., Zheng, B., Zhang, S. et al. (2020). Intelligent reflecting surface meets OFDM: protocol design and rate maximization. *IEEE Transactions on Communications* 68 (7): 4522–4535, doi: 10.1109/TCOMM.2020. 2981458, https://doi.org/10.1109/TCOMM.2020.2981458.

28 Bose, A. and Foh, C.H. (2007). A practical path loss model for indoor WiFi positioning enhancement. *2007 6th International Conference on Information, Communications & Signal Processing*. IEEE, pp. 1–5.

29 Özdogan, O., Björnson, E., and Larsson, E.G. (2019). Intelligent reflecting surfaces: physics, propagation, and pathloss modeling. *IEEE Wireless Communications Letters* 9 (5): 581–585.

30 Kisseleff, S., Martins, W.A., Al-Hraishawi, H. et al. (2020). Reconfigurable intelligent surfaces for smart cities: research challenges and opportunities. *IEEE Open Journal of the Communications Society* 1: 1781–1797.

9

Performance Analysis of UAV-Enabled Disaster Recovery Networks

Rabeea Basir[1,5], Saad Qaisar[1,2], Mudassar Ali[1,3], Naveed Ahmad Chughtai[4], Muhammad Ali Imran[5], and Anas Hashmi[2]

[1] School of Electrical Engineering and Computer Science (SEECS), National University of Sciences and Technology, Islamabad, Pakistan
[2] Department of Electrical and Electronic Engineering, University of Jeddah, Jeddah, KSA
[3] Department of Telecommunication Engineering, University of Engineering and Technology, Taxila, Pakistan
[4] Military College of Signals, National University of Sciences and Technology, Rawalpindi, Pakistan
[5] James Watt South Building, School of Engineering, University of Glasgow, UK

9.1   Introduction

In 2014, there is a record in the Annual Statistical review of 324 number of disasters that occurred around the globe, leaving around 8 million victims and 140 million affected human beings [1]. The communication infrastructure (cellular network) is usually distorted after a disaster, and it is very important to have updated information for rescuing the disturbing area. Previously, mobile ad hoc networks (MANETs) were used to provide services [2–4], which takes more time in deploying an ad hoc network. They can deploy a multi-hop network, which provides communication. Similarly, vehicular ad hoc networks (VANETs) could provide post-disaster recovery services, but for the transport of vehicles, roads may be severely destructed. Similarly, for satellite communication, channels are not distinctive and may be distorted after the discovery. It is not possible to deploy a new network or manage the already deployed infrastructure for a disaster-affected area in a very short time. However, the development of small UAVs/drones/FR has paved the way for providing communication solutions for critical environments (military or disaster rescue missions) [5–7]. UAVs can reach almost every location, crossing the obstacles in a disaster-affected area.

Autonomous Airborne Wireless Networks, First Edition.
Edited by Muhammad Ali Imran, Oluwakayode Onireti, Shuja Ansari, and Qammer H. Abbasi.
© 2021 John Wiley & Sons Ltd. Published 2021 by John Wiley & Sons Ltd.

They are also named aerial ad hoc networks (AANET) [6] or flying ad hoc networks (FANET) [8]. Many factors affect the performance of unmanned aerial vehicle (UAV)-enabled disaster recovery networks; Section 9.4 provides details of all such factors in detail. Researchers could provide performance-efficient solutions, keeping in view a particular objective.

9.2 UAV Networks

From the last decade, UAV networks have been studied and observed by the researchers. They have supported the wireless network's utility with their promising attributes of high reliability, high intelligence, enormous mobility, and cost-effectiveness. UAVs in connection with existing cellular wireless networks have maximized the data rate production, improved the coverage capability, and increased the network capacity. The hard and high-performance requirements of future 5G heterogeneous networks could be fulfilled by UAVs supporting different applications. Based on many interactive benefits, it is estimated that by the year 2025, there will be a demand for UAVs/drones/FR worth about $45.8 billion for commercial and industrial use cases [9].

The enormous generation of data requires the deployment of more base stations to satisfy the demands of users, but the deployment of new infrastructure is not a cost-effective solution to this coverage problem. UAVs can solve this issue without the amendments in the existing network infrastructure. Researchers have provided some solutions to maximize the capacity and coverage of mobile communication systems. Their dynamic and flexible support could manage and accommodate the enormous number of increasing users in the network, with a view to maximizing the throughput coverage. They can cater to problems of high demand and overloaded communication in networks. Because of the high mobility feature, they can help in the realization of emergencies.

When a disaster occurs, all the existing communication goes down or the network gets damaged. One of the key advantages of UAVs is that they can provide communication services during disaster situations or in emergency environments. Cache-enabled unmanned aerial vehicle-base stations (UAV-BSs) have also gained attention as a possible solution to provide reliable and nonstop connectivity. They can work in the failure of terrestrial base station (BS), which could be the result of any disaster situation. In this case, UAV-BSs can be rapidly deployed to provide communication. Deployment of cache-enabled UAV-BSs in association with Internet-of-things (IoT) networks can reduce the latency requirement and also provide energy-efficient communication. UAV-aided communication improves coordination, takes intelligent decisions, and controls traffic regulations.

The UAV network is a combination of MANET and VANET, as there is a function of both networks in UAV network. Unlike MANETs, they have high mobility; similarly, unlike VANETs, UAVs move very dynamically, which results in a continuous change of topology. Different types of architectures enable different applications. A single UAV system comprises a single unit that has to work and communicate with the ground node as an isolated node. This single UAV system is also characterized by another name, UAV-to-BS; they provide more effective, accurate, and reliable results. Rather than this one large UAV node, they are more beneficial when they act as a group. The multi-UAV systems, being more powerful and capable of doing more tasks, provide solutions for the monitoring of a hazardous environment. Because of group tasks, these can be named UAV-to-UAV (U2U) systems. Multi-UAV systems have many benefits: such missions can be completed cost-effectively; their collaborative work provides constant services, which improves the network performance; the task could be done at any cost even if one UAV fails to work; missions could be time-saving.

Despite all these benefits, the deployment of UAVs in different existing networks faces many problems. Deployment, cooperation, positioning, and mapping of UAVs with existing network architecture are major issues. Handling a large number of users in different applications, demanding different quality of service (QoS) requirements, and load balancing are challenges that need to be solved. Some major challenges for UAV-aided communication are 3D placement problems, resource allocation problems, and sensor deployment problems on the basis of some specific objective. The 3D placement problem also causes the coverage problem as it is no longer a 2D deployment. Placement depends on three factors of height, altitude, and longitude. Resource allocation could be done on the basis of maximum capacity, maximum energy efficiency, minimum cost, minimum latency, and minimum power consumption. Sensor deployment problem is one major challenge to be solved; in a multi-UAV system, the deployment on different sensors might be different to collect specific data. For efficient performance, the communication and networking issues (noisy channels, collisions, coverage range, latency, and limited spectrum) should be solved to enable and organize multiple UAVs.

9.2.1 UAV System's Architecture

Based on networking, there are different architectures of UAV systems as shown in Figures 9.1 and 9.2. They are single UAV, multi-UAV, and

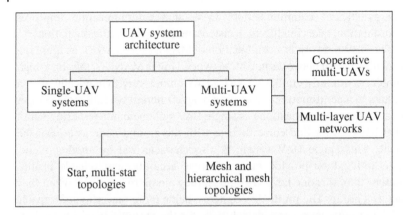

Figure 9.1 Flow chart explaining different architectures of UAV systems.

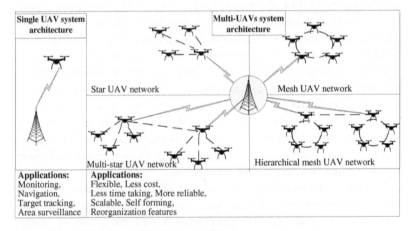

Figure 9.2 UAV system's different topologies.

cooperative multi-UAV systems. Depending on the service requirement, the best possible architecture is chosen in such a way that it increases the performance of the network.

9.2.1.1 Single UAV Systems

In this system architecture, there is a direct connection between the UAV node and the base station. It does not mean that there is only one UAV node in the system; there might be more than one UAV node to provide the service. Many UAVs in the system enable efficient performance in the network. The performance of UAV only depends on the UAV-to-base station communication link. For distant targets in an unknown area, such architecture is used for monitoring and controlling services. A single UAV system can also be used where the operational area or weather or flight conditions are critical and it is not safe to send a plane. More than one UAV node in the system acts like an isolated node that provides different services.

From a disaster recovery network point of view, single UAV systems can be used to track the vehicle or to move the vehicle to the desired location; there is still an open issue of the limited capability of a single-UAV node in the system to perform the task, for which some researchers provide solutions while optimizing the sensor's functionality, hardware, and software. Multiple objects can be observed using a single UAV system; it helps in the fast localization of objects. It also enables to investigate the exact size of the operational area that was disturbed after a disaster. Navigation, monitoring, and control can be done using a single-UAV system. It also saves the use of extra resources to perform a task.

9.2.1.2 Multi-UAV Systems

Different network topologies establish the multi-UAV system. All network topologies, namely star, multi-star, mesh, and hierarchical mesh, bring more efficient results for critical applications such as disaster management and military missions. For a star topology, the performance is only dependent upon the UAV-to-BS communication link. The base ground station is attached directly to a UAV node. For multi-star topology, more than one group of UAVs in a star topology are present in the system. One UAV node from every star has a direct connection with the base station. These nodes also have an inter-UAV communication link with another star's node. There is a need for more bandwidth for this topology, which also adds more delays in terms of propagation delay between the links. Because of this unwanted propagation, energy will also be wasted. From the name, it is clear that in a mesh topology, UAVs are connected among themselves and only one UAV has a direct connection with the central BS. Similar to the multi-star topology, in hierarchical topology, there are a greater number of meshes present in the system. Every mesh has a node that has a direct communication link with the ground BS. In comparison with a star topology, mesh topology gives more flexibility and provides more reliable performance. The next two topologies are the extension of the multi-UAV system.

9.2.1.3 Cooperative Multi-UAVs

For applications where more than one service is required at one time, cooperative multi-UAV systems are installed. Their performance depends either on task completion or on time threshold. In the cooperative system, the

UAV nodes form a topology in form of a graph. The task-based cooperative multi-UAVs provide services for both static and dynamic environments. In a static environment the functionalities are almost constant, while in a dynamic environment there will be a frequent topology change that requires different services. In this system, coordination among the UAVs is required to perform a distributed complex task. Intelligent programming for task distribution provides better solutions, more specifically for dynamic environment applications. The dynamic environment results in more frequent links failure; for completing a mission or task, opportunistic relaying is required, which enhances the usage of network resources. For the reconstruction of the disaster-affected area, delay-tolerant graph topology of UAVs is required. It will work on the objective function of minimizing the latency.

9.2.1.4 Multilayer UAV Networks
In coordination with other architectures (IoT/wireless sensor network (WSN)/cellular/cloud computing), UAVs form a layered network topology. This will provide better performance results.

- Future **smart IoT applications** (transport, mobility, healthcare, etc.) can be efficiently performed using UAV networks. Internet of drones (IoD) is a term used for the collaboration of UAVs with IoT technology. This provides services of search, control, and rescue.
- **Cloud computing** association is used to overcome the limited computing power and the storage capacity of UAVs in the network. This will reduce delays and enhance performance in terms of less energy consumption and data processing.
- UAVs in form of groups **(swarms)** will help in completing the mission when the operational area is very large. They also require an efficient and secure communication link for coordination among themselves. This UAV swarm's formation helps in completing complex missions, keeping in view the collision avoidance issues. The distribution of tasks is also optimized in a group (clusters) of UAV nodes, as UAV formed clusters. Only the cluster head has to connect with the central station, which ensures a constant connection. This group formation also enhances the sensing capabilities and increases the performance of the network.
- **Sensors** on the UAV nodes enable the monitoring and investigation parameters such as temperature, wind, and air temperature. In a disaster recovery network, UAVs along with aerial sensor networks collect data from different areas. Multiple UAVs with different types of cameras installed gather the information in the form of images or videos. As an

example, two different types of cameras, high-resolution and infrared cameras, are used to collect information over the operational area. For this ground sensor network and UAVs, the sensing network must have the coordination to perform well.

For disaster management and emergency situations, multi-UAV systems are used to provide services. In this system architecture, UAV nodes are smaller in size and perform tasks in an efficient, coordinated way. They have the capability of self-forming and reorganization, if a node fails to work other UAVs will do the reconfiguration. This system requires both communication links (UAV-to-BS and U2U) for better performance in the network. The performance of a UAV node can be monitored via the base station if all UAVs have a direct connection with the infrastructure ground station. This connection brings a constraint of the coverage area, for example, in a dynamic environment condition such as the disaster-affected area; the UAV might get disconnected because of random node movement and disturbed area. In comparison with a single UAV system, the multi-UAV systems bring many benefits that improve the overall network performance. Multi-UAV system gives reliable results; the mission can be completed in less time and with less resource utilization; more scalable results can be found in recovering a disaster-affected area; multi-UAV systems can provide services over long-range areas and a more survivable network, for example, if a UAV node fails to perform the task, the task can be completed by the coordination of the other UAV nodes in the system.

9.3 Benefits of UAV Networks

In this section, the most prominent UAV applications are discussed. UAVs are used in many ways such as agricultural, military, commercial, and environmental applications, namely estimation of wind, monitoring of traffic, sensing, wildfire management, surveillance of border, monitoring of disaster, operations of search and destroy, and insurance sector, healthcare sector, educational sector, transportation, and entertainment sector [10–14]. All applications are different as they have different system requirements in terms of varying parameters such as delay tolerance, frequency, energy limitations, and coverage range. These parameters directly affect the performance of the network. Figure 9.3 demonstrates the different benefits of UAV networks.

1. *Healthcare*: Smart healthcare is one promising application of future net-works, in which the patient does not have to visit for a checkup every

Figure 9.3 UAV system's benefits in different applications.

time. For collecting information in a disaster recovery network about the victims affected by the natural disaster (earthquake, floods, etc.) or infectious disease (coronavirus infection), UAVs are used. These UAVs are organized with artificial intelligence algorithms and hardware sensors. For providing health services in the affected disaster area, UAVs might send some aid to the victims. Similarly, for such emergency situations items of everyday use could be sent using drones.

2. *Target tracking or localization*: For military applications, localization and tracking of the target are important parameters. UAVs or drones can fly over such areas and could calculate the exact location of the target. Many researchers have already provided solutions using the UAV network as an enabler of many applications. Similarly, for disaster-affected areas (fire, flood, etc.), cameras and sensors on the UAV could capture information regarding the exact location of the environment's disturbing point. UAVs also help in real-time vehicle detection in case of a road accident. The localization of the target depends on the position and mobility of UAVs. The tracking algorithm works based on past operation calculations. For example, for disaster management applications, it is helpful to use the previous information to track the changes. This knowledge might help in the fast recovery process.

3. *Search and rescue (SAR)*: Owing to the flexible and dynamic nature of UAVs, UAVs can be used to provide services for emergencies such as

natural disasters, road accidents, and situations where urgent evacuation is needed (fire, flood, and storm). These types of applications require low latency processing and high speed. Different architecture types of UAVs (single, multi, and cooperative) can be used depending on the crisis. To quantify such applications, the UAV's task distribution in the operational area is an important challenge to cater.

4. *Area surveillance*: Investigation over the operational area (natural disaster, emergencies) is required more frequently. A group of UAVs is required for perfect area surveillance. For the efficient investigation of the area, the best type of sensors is required to work over the dynamic environment. Similarly, large computing power helps in more area surveillance in less time.

5. *Data collection*: To recover any type of network after a natural disaster over different areas, such as deserts, mountains, and plane areas, UAVs are used to gather the updates. WSNs along with UAV nodes enable the collection of data over the whole operational area. The decentralized nature of UAVs and the dynamic nature of the operational area make it difficult to collect all updates, for which optimized solutions are required, which considers low latency and low energy consumption factors.

6. *Inspection*: After a disaster crisis, there are some unreachable parts in the operational area that are not easily accessible by the operators. UAVs can inspect and navigate all portions in very little time. UAV, in association with other technologies such as computer vision, artificial intelligence, and machine learning, is used for monitoring railway and airport infrastructure. For the recovery process in a disturbing sector, the deployment of UAV provides robust monitoring, controlling, and inspection.

7. *Construction*: Owing to the installation of a high-resolution camera on UAVs, they can be used for improving the performance of safety construction. There are two existing approaches to construct safety in the network. The first one identifies the issues related to safety, while the second one generates routes for the UAV node's flight followed by data collection in the form of images by cameras. Safety measures can be done during the reconstruction of the operational area. For better reconstruction of a network after the emergency crisis, UAVs could take videos and images constantly. Experts can study the current situation of all unsafe paths, unsecured material, unprotected edges, and damaged paths.

8. *Transportation*: Smart transportation sector takes UAVs as an enabler of delivering the packets more speedily. For a disaster recovery network, when no roads or transportation system is available for normal movement and delivery of goods, UAV provides these services. UAVs also

reduce the congestion, delivery time, and traffic that are experienced on the ground. UAVs provide more reliable and fast transportation among cities.

9.4 Design Consideration of UAV Networks

To gain the benefits of UAV networks, there are many system considerations. This section presents the most prominent design requirements of the UAV network.

1. *Reliability*: The performance of the network can be degraded under the effect of some network attacks. Reliability is one of the basic requirements of UAV networks. A reliable network is also fault-tolerant and fast. Reliability is an application-based optimization factor for UAV networks. For example, videos and pictures can be sent using low reliability, while a disaster situation requires robust and highly reliable communication. High reliability can be gained by low latency and high energy-efficient communication. Recovery algorithms, pre-analysis of the network, and latency-tolerant functionalities are used to give high reliability. The random movement of UAV nodes in a disaster situation might cause a less reliable network, which is why it is one important factor that should be under consideration for designing a UAV-enabled disaster recovery network.

2. *Mobility*: Mobility is one important factor for designing UAV networks, as such networks are characterized by high mobility factor and highly mobile nodes. This mobility factor along with different speeds of mobile nodes changes the topology of the network more frequently. The network gets partitioned every time depending on the application use case. For example, for a disaster situation such as an earthquake or storm, UAVs will fly over the affected area to provide communication over slow or dynamic links, while for the agriculture sector UAVs will fly over large-scaled areas and communication links are more dynamic. For a disaster recovery network, UAVs could serve in a random manner or a group of two or three, i.e. in an organized manner. The organized way named as swarms (in a group of 2) involves rapid movement in three different dimensions. In some applications the mobility is not random; UAVs probably visit some known places such as over an agricultural or forest field. For a disaster area of operation such as an earthquake, storming, or forest fire, UAVs have to move over different areas at different speeds, handling different burdens. The visiting probability of a

UAV increases the probability of packet delivery, which normalizes the disaster situation. In a disaster recovery network, UAVs have to move with the highest probability. Every time this high mobility changes the topology of the network, it affects the relative positions of UAVs. This might fail a UAV or a communication link; the other relative UAV node must complete the operation in the disaster area without disrupting the user requirement. This node arrangement and resource management under high mobility factor are one of the major requirements that affect the network performance.

3. *Scalability*: For improving the performance of a disaster recovery network, scalability is an important factor. It also minimizes the cost in terms of latency, and energy usage during re-planning and reorganizing of the network. The network should be flexible to such an extent that it allows the entrance of new UAVs at any time. The scalability of the UAV network is related to the number of resources. For a UAV-enabled disaster recovery network, latency and energy management are important parameters. To recover the operational area in less time and in a more energy-efficient manner, the UAV network should be highly scalable. Other than the disaster effects in the operational area, there might be obstacles such as mountains, walls, and buildings, which can influence the communication coverage and may stop the radio signals between the UAVs and base stations. To tackle this, UAV-enabled disaster recovery network should be highly scalable to improve performance. Scalability can be increased by optimizing the number of relay nodes in the network.

4. *Adaptability*: UAV placement, the distance between UAVs and base stations, and routes are major changes in UAV networks that are deployed in a disaster sector. The mission may not be fulfilled because of the dynamic nature of UAVs in the disaster network. These changes affect the performance analysis of the network in terms of latency, recovery, energy, and cost parameters. In a disaster network, the UAV might get frequent failures, or the number of UAVs to recover the operational area is not enough, or weather conditions might affect the communication links. All such conditions change the performance of UAVs and affect the mission. For such applications, the UAV system must be adjustable and adaptable by itself to cater to the mentioned failure conditions or requirements.

5. *Bandwidth*: For a disaster recovery network there is a requirement of high data rate transfer; hence more bandwidth is required. Spectrum allocation is dependent upon the application; UAV applications in normal conditions do not require more bandwidth as the transfer of data is

not high whereas for disaster conditions such as earthquakes or storms more data rate transfer is required for more updated information. Spectrum allocation is based on the efficiency of energy and bandwidth. Frequent changes in the UAV network require more bandwidth, and bandwidth is dependent on the channel capacity, speed of UAVs, and error-prone threshold of the wireless links. UAVs along with sensor technology have to collect more data, which eventually requires more bandwidth. For secure communication, efficient allocation of spectrum is required. For UAV-enabled disaster recovery networks, high and efficient bandwidth allocation is required, as a high data rate is needed. More bandwidth increases the performance of UAV networks.

6. *Latency*: Low latency is a major requirement for all future applications. For disaster recovery networks it is one of the greater parameters for consideration. UAV networks must reduce all types of delays. Delays are controlled on the basis of the application; for example, for applications that are time-critical, such as disaster situation and military applications, overall delays must be very less. Such optimized latency minimization solutions need to be proposed for disaster recovery applications. For UAVs, resource management issues need to be observed with an objective of less latency. In a multi-UAV network, there are collision problems that cause delays, which must be catered for disaster recovery applications.

7. *Power consumption*: Power consumption of UAVs in a time-critical environment is another important parameter. For small UAVs, power is drained more quickly as compared to the large UAVs. This difference will affect the overall performance of the disaster recovery network, which eventually requires more time to recover. Power consumption and energy efficiency of the network are related. Similarly, if less power is consumed, the network lifetime can be enhanced. For a disaster operational area, power and latency are major parameters that need to be discussed. Optimized power-controlled solutions need to be developed for enhancing network performance.

8. *Security*: The UAV networks' composition has made them very vulnerable to external attacks. They are structured like a cyber-physical system (CPS), having many components such as control, communication, computation and sensors, actuators, and communication equipment. UAVs are autopilot devices and are equipped with sensor technology such as vision, radar, infrared sensors, and global positioning system (GPS) receivers. For information transfer, navigation, and guidance, UAVs are dependent upon these sensors. For a disaster situation, the corrupted sensor can cause serious issues. Instead of recovering the

operational area, this may affect the network badly. Sensor spoofing attacks disturb UAV mode configuration, which affects the performance of the UAV network in recovering from the disaster area. Fault-information transfer over GPS channels and poor vision sensors cause a decline in the performance of the recovery network. This also causes more delay in recovering the operational area. This requirement of error-free communication might result in more power usage.

During the recovery process of the affected area, the wireless links are prone to many attacks such as misuse of information, eavesdropping, interference from other UAVs, denial of service, and the tampering of sensitive information. Similarly, the random dynamic topology of the UAV network results in difficulty in finding a disturbing node in the network, which declines the performance.

Sensors on UAV nodes can be controlled by an attacker. The attacker might take control of the flight of UAV nodes, which eventually lowers the performance during recovery of the disaster operational area. Moreover, in a natural disaster situation such as earthquakes, storms, floods, and cyclones, the environment is uncontrolled, which might enhance the external attacks. Within a UAV transmission range, an attacker can send and receive information at any time, altering the actual information. In such situations, for the route discovery, the broadcast message can be received by the wrong receivers. Or maybe the selected route is more prone to errors. UAVs mostly have limited computation power and storage capacity. Computation is done on the available data for decision making. An attacker can drain all the battery of the UAV network using sleep deprivation. Similarly, a flooding attack can exhaust the available network bandwidth and consumption of all UAVs. For disaster situations, the available limited bandwidth and power are very important parameters that have effects on network performance. For a UAV-enabled disaster recovery network, all these factors cause security problems. External attacks on any mentioned component of the network can cause serious issues. For efficient performance of the UAV network, all parameters must be considered during development.

9. *Privacy*: This parameter is related to the previously mentioned security factor. For all applications, UAVs can capture videos and collect information. This feature may result is a crucial issue for some applications such as military or accident situations; such sensitive information can be encrypted by any outsider. Therefore, for improving the network performance privacy should be maintained and new techniques are required to guarantee security and privacy.

10. *Localization*: For a disaster recovery network, the exact location of UAV nodes is a very critical point to study. It is more important when more than one node is present in the network. It is difficult to find the exact location of the UAV on the basis of approximation. GPS technology was an existing technology used for finding the location of UAV nodes, but it is not effective for some sensitive and critical applications such as earthquakes and military missions. For such an application, UAVs must have a fast speed, which is characterized by knowing the exact position in the network. This localization information must be exchanged over small time intervals to improve network performance in terms of latency, energy consumption, and privacy. Many authors have suggested solutions other than GPS for the localization of UAVs.

11. *UAV platform constraints*: The two important constraints are weight and space limitation. Weight is related to hardware, while space limitation is for mini-size UAVs. Payloads are lighter when the UAV hardware is light in weight, while a greater number of sensors can be implanted over the surface when the hardware is heavier. The material of UAV also affects the performance of the network. Similarly, the space limitation of small-sized UAVs also has effects on communication. For applications such as earthquakes and floods, such platforms that seem not important at first must be taken into consideration.

12. *Coverage* [5]: Every UAV node has a limited coverage parameter. Coverage factor is of two types, namely network coverage and area coverage. Network coverage involves UAVs acting like communication relays that provide connectivity between the receiver and the base station, whereas, in area coverage UAVs are used to collect information related to maps, monitoring the area, and investigation. Coverage is one major factor that affects the performance of UAVs in a disaster recovery network. More coverage by UAV nodes will result in less usage of resources; otherwise, if coverage is small, a greater number of UAV nodes are required to cover the entire network. The coverage factor of UAVs for critical applications as for the unstable and dynamic environment is affected by many other factors such as energy consumption, size of the operational area, obstacles present in the operational area, and speed of UAVs. The high mobility of UAV nodes in the network has a direct relation with the coverage. More speed of UAVs results in early coverage of the operational area, while this results in the complexity of algorithms. Similarly, energy has direct effects on the coverage capability of UAV nodes. If the energy stored in the node is less, the UAV cannot cover the operational area and there will be early drainage of energy. The coverage

factor of a UAV node in an operational area of disaster recovery network must be able to consider all the obstacles such as mountains, buildings, and trees. UAV nodes management, which chooses an optimized collision-free and safe route for recovering, is needed. The coverage of UAV nodes is dependent upon the application type - if the application required static UAV nodes, coverage is static, while for critical applications, UAVs are flying and their coverage is also dynamic. The coverage of UAV nodes in a disaster recovery network is an important parameter that affects the performance in terms of some factors mentioned in this subsection.

13. *Topology*: To complete a task, either a single cluster or multi-cluster of UAV nodes is formed. This topology formation helps in maintaining coordination and collaboration. For disaster recovery networks, UAV nodes must have multi-cluster formation, which helps in enhancing the network performance. For high coverage and recovery in less time, an optimized topology of UAVs is required. In multi-cluster network formation, the cluster head of each cluster does the communication. These cluster heads are used for downlink communication.

14. **Prediction**: For recovery of a network after a disaster, movement and position of UAV nodes are important. The position of UAV is dependent on various factors of speed, direction, operational area, and mobility model. It is very difficult to predict the performance of the network based on these factors.

9.5 New Technology and Infrastructure Trends

There are positive effects on the performance of the UAV systems with the integration of other infrastructures such as network function virtualization (NFV), software-defined networks (SDNs), cloud computing, image processing, and millimeter wave (mmWave) communication. Similarly, enhancement in technologies such as artificial intelligence, machine learning, optimization theory, and game theory improves the overall performance of the network. The effects of other infrastructure and technologies are reviewed under this section. All the mentioned contributions are summarized in form of a table as well (Table 9.1). To gain an idea about the diverse use of UAV system in different applications, a comparative analysis is given in the table. The comparison is based on different factors, which affect the performance of a network. Further, we discuss existing solutions to the integration of architectures with UAV.

Table 9.1 Critical review on state of the art.

References	Approach	Application	Performance effecting factors	Problem
[15]	UAV allocation algorithm	UAV vehicles	Security	Optimizing the available resources
[16]	Placement scheme for MEC-NFV framework		Reliability, bandwidth/spectrum, latency, cost	Optimal placement and provisioning of UAVs' services
[45]	Optimal drone scheduling algorithm		Reliability, energy	Optimal scheduling of UAVs
[46]	Employed container-based monitoring and anomaly detection network functions (NFs)	Telemetry monitoring (80 flights)	Latency	Increasing situational awareness, reducing the latency of monitoring and burden on the network
[47]	Cost-effective solution is provided by the implementation of VNFs over the infrastructure of SUAVs		Latency	Increasing the performance of network services
[17]	UAV-assisted vehicular computation cost optimization (UVCO) algorithm	Vehicular network	Latency, energy, cost	UAV-assisted vehicular computation offloading decision-making problem
[20]	Comparative analysis with or without SDN integration with UAV network	Monitoring sensors	Bandwidth/Spectrum	Bandwidth and packet loss distribution
[48]	Ortho-mosaic and digital terrain model (DTM) generation	Landslide investigation	Cost	Low-cost monitoring of landsliding
[26]	Hybrid image binarization	Wall crack identification		Identifying crack width accurately while minimizing the loss of crack length

Ref	Method	Application	Metrics	Objective
[18]	Joint mode selection and resource allocation optimization (JMSRAO) algorithm		Latency, energy	Minimizing the weighted sum of the delay and energy consumption
[21]	SDN and MQTT hybrid structure + QoS-based multipath routing framework	Battlefield UAV swarms and two case studies to validate the infrastructure	Bandwidth/Spectrum, security	Network performance enhancement
[49]	Load balancing algorithm		Bandwidth/Spectrum, latency, routing	Routing and link switching by ensuring load balance for end-to-end data delivery in a dynamic UAV network
[19]	SDN based topology management for FANETs (construction, adjustment, Integration, and node allocation)		Reliability, latency, routing	To propose a novel coordination protocol, which includes an SDN-based UAV communication for routing and topology management
[25]	Open Jackson network theory + Queuing theory	Military and civilian applications	Reliability	To find out how much data is generated and processed by UAV sensors while keeping the system stable and reliable
[24]	Markov approximation algorithm	A case study for latency-critical applications in smart cities	Bandwidth/Spectrum, routing	Joint task placement and routing (JTPR) problem
[22]	Markov approximation technique (MA algorithm) + Lyapunov optimization	Long-term high-efficiency and stable performance in an online environment	Latency, cost	Jointly optimizing workflow assignment and multi-hop traffic routing (JOAR) problem

(Continued)

Table 9.1 (Continued)

References	Approach	Application	Performance effecting factors	Problem
[23]	Two heuristic algorithms		Security, cost	Joint resource allocation and computation offloading problem
[50]	New framework layout of fog computing with UAV network is developed	Large-scale mission and search operations	Bandwidth/Spectrum, latency, energy	To overcome the throughput and latency involving multiple aircraft
[51]	A system for visual recognition is proposed for industrial internet of things (IIoT)-based UAV to calculate the latency	A case study in an industrial concrete plant production line	Latency	To provide real-time visualization of the plant and study three-layered latency framework
[52]	Three technologies, namely the normalized difference vegetation index (NDVI), the near-infrared spectroscopy (NIRS), and the digital elevation model (DEM), are used to observe the contribution	A case study to observe an agriculture framework		To analyze the agriculture using mounted remote-sensing drones
[27]	Hat transform and HSV threshold method	Health monitoring (crack detection and assessment of surface degradation) of civil structures	Latency, cost	To determine primary structural defects
[28]	ThingSpeak cloud application	Forest fire detection system in smart cities	Latency	To detect fire in a forest in real-time and sending warning alerts via emails
[53]	Morphological operations on images	Intelligent transportation system		Automatic detection of potholes

Ref	Algorithm	Deployment	Parameters	Goal
[31]	A greedy user scheduling algorithm		Reliability, bandwidth/Spectrum	To decrease the blockage probability and maximize the achievable sum rates of the multi-UAVs system
[32]	Proposed hybrid design method + the geometric greedy algorithm		Accuracy	To obtain precise training beams in the 3D scenarios
[54]	The ray tracing (RT) theory in traditional geometry-based stochastic model (GBSM)	Deployment in campus scenario (air base station layout and cruise routes)	Latency, power	To develop a system that studies performance optimization and evaluation of UAV mmWave communication systems
[33]	Lagrangian dual decomposition method		Power, capacity/rate/throughput	Resource allocation problem to maximize the sum rate
[55]	An iterative algorithm		Power, security	Maximizes the minimum secrecy rate
[56]	Coalitional game + Lyapunov optimization		Mobility, latency, cost	To maximize network payoff (cost and delay)
[57]	Iteratively optimizing two convex sub-problems		Capacity/Rate/Throughput	To maximize throughput (three-dimensional location and spectrum sensing duration)
[58]	Iterative algorithm (variable substitution, successive convex optimization techniques, and the block coordinate descent algorithm)		Power, capacity/rate/throughput	Joint resource allocation and UAV trajectory optimization
[59]	Iterative algorithm based on successive convex approximation (SCA)		Power, security	To maximize the secrecy rate

(Continued)

Table 9.1 (Continued)

References	Approach	Application	Performance effecting factors	Problem
[60]	Optimal power allocation scheme + fractional programming (FP)	Disasters scenarios (emergency areas, wide areas, and dense areas)	Bandwidth/Spectrum, power	Joint UAV deployment and resource allocation (the user scheduling, the mobility of the UAVs and transmit power control)
[37]	A decaying deep Q-network (D-DQN)-based algorithm		Bandwidth/Spectrum, power	To minimize energy consumption
[38]	Liquid state machine (LSM)-based prediction algorithm		Bandwidth/Spectrum, energy	To maximize the number of users in a stable queue
[61]	Three-step machine-learning-based UAV placement (partitioning, placement, real-time movement)		Power	Joint power allocation and trajectory design
[39]	A three-step approach (Q-learning-based placement algorithm, an echo state network-based prediction algorithm, a multi-agent Q-learning-based trajectory acquisition and power control algorithm)		Mobility, power, capacity/rate/throughput	To maximize the data rate
[62]	Iterative sequential minimal optimization (SMO) training algorithm			To achieve an efficient and low-complexity code word selection

Ref	Technique	Application	Metrics	Objective
[63]	The weighted expectation maximization (WEM) algorithm		Latency, energy, capacity/rate/throughput	To maximize the utility of UAV network (deployment)
[64]	Artificial neural network (ANN)-based solution schemes	Delivery systems, real-time multimedia streaming networks, and intelligent transportation systems	Reliability, security	To highlight wireless and security issues
[65]	Big-data-aided feature extraction and ML-aided optimization solutions		Reliability	To enhance the service quality experienced by users
[34]	Swarm-intelligence-based localization (SIL) and clustering schemes	Emergency communications	Mobility, latency, routing, energy	To design a topology for UAV network, which supports high mobility and caters to dynamic nature
[35]	Genetic algorithm (GA) and simulated annealing (SA) algorithms	Danger area	Bandwidth/Spectrum, latency, capacity/rate/throughput	To determine the optimum number of drones and their optimum location
[36]	A swarm intelligence-inspired autonomous flocking control scheme (SIMFC)	Simulator-based testing on OMNeT++	Energy	To maintain the topology of UAVs during flying process keeping in view the quality of service
[66]	A distributed algorithm		Latency, energy, cost	To maximize the utility function

(Continued)

Table 9.1 (Continued)

References	Approach	Application	Performance effecting factors	Problem
[43]	Evolutionary equilibrium (EE) + Nash equilibrium (NE)		Bandwidth/Spectrum, cost	Joint access selection and bandwidth allocation
[67]	An efficient schemer based on software defined network (SDN) with media independent handover (MIH)		Mobility, latency, capacity/rate/throughput	To maximize the utility (end-to-end delay + handover latency + signaling overheads)
[41]	The multi-UAV energy-efficient coverage deployment algorithm based on spatial adaptive play (MUECD-SAP)		Reliability, power, energy	Coverage maximization and power control
[44]	Distributed computation offloading		Latency, power, cost	Cost minimization (energy and delay)
[42]	Artificial intelligence (AI) and game theory (GT)-based near real-time flight control algorithms	OPNET modeler environment		To observe the connectivity and uniform distribution of swarms over MANET

9.5.1 Network Function Virtualization (NFV)

In [15], the authors have proposed virtual network security functions (VSFs) to enhance end-to-end security in multi-access edge computing (MEC)-UAV system. They have analyzed different kinds of functions, VSFs, namely virtual firewalls, virtual intrusion detection, and virtual authentication authorization accounting, on real drones to observe the performance. Two use cases are also summarized where UAVs are deployed as a mobile node. The proposed problem's efficiency is calculated in terms of the time it takes to deploy a new VNF on an SDN configuration. UAV node and flying time performance are observed in the real UAV deployment.

To design low-cost UAVs traffic management (UTM) systems, an integer linear programming problem is formulated to minimize the total deployment cost under the node's placement, node's capacity, latency, reliability, and routing constraints [16]. NFV is integrated with MEC-UAV to provide QoS-aware Management and Orchestration (MANO) services for the proposed cloud and MEC platforms.

9.5.2 Software-Defined Networks (SDNs)

SDNs have shown potential in data collection and centralized management application. SDNs can enhance UAV networks by providing flexibility, adaptability, and centralized control. In [17], average system cost (ASC) is minimized by proposing UAV-assisted vehicular computation cost optimization (UVCO) algorithm. SDN is associated with MEC and UAV architecture to optimize the system cost of vehicle computing tasks. SDN gives support in terms of data collection and management, while MEC enables low latency communication. To optimize, the UAV is used as an intelligent relay node to transfer the task from the vehicle user to the MEC server. Low complexity and decentralized solutions are found using game theory. A mixed-integer combined non-convex minimization problem is formulated in [18]. The main objective is to minimize the weighted sum of delay and energy consumption of UAV-assisted cellular users, under mode selection and resource allocation constraints. For the solution of a joint mode selection and resource allocation, an optimization algorithm is proposed, which works in two steps, namely, branch and bound method and convex optimization method. The performance of the proposed algorithm is compared with a random mode selection scheme. e Silva et al. [19] proposes an SDN-based topology management for UAV networks. The authors establish FANETs based on SDN to provide continuous and reliable communication links among UAVs. Vishnevsky et al. [20] studied SDN-assisted UAV networks for monitoring sensor data.

In [21], the authors proposed SDN and message queuing telemetry transport (MQTT)-based hybrid architecture for battlefield UAV swarms. The authors also proposed a QoS-based multipath routing framework.

9.5.3 Cloud Computing

In [22], problem of a joint optimization for UAV swarms that considers computation offloading and multi-hop routing schedule is formulated. In both peacetime and emergency situations, UAV swarms are a very useful solution for topographic mapping and traffic control. The authors tackle the NP-hard problem by developing a Markov approximation-based method to find the near-optimal solution. Markov approximation and Lyapunov optimization are jointly used to optimize the cost of online and highly dynamic environment for long-term performance. The results indicate lower routing costs, which are based on low computations. Khan et al. [23] evaluate an energy consumption minimization problem for resource allocation and computation offloading strategy in a UAV-enabled secure edge cloud computing system. A heuristic algorithm is proposed to solve the problem by dividing it into three sub-problems. Resource allocation, task partition, and computation offloading are optimized energy efficiently using the proposed solution. The results show a comparison of the proposed algorithm with related schemes.

For resource-intensive applications such as crowd sensing for smart cities and real-time monitoring at the time of protests, UAV-edge-cloud computing model is studied in [24]. A hybrid computing model is proposed to achieve high QoS. A joint task assignment and routing problems are studied for latency critical application. Luo et al. [25] address the issue that UAV networks cannot handle the big data generated by the large number of sensors that may monitor some disaster area. By merging the UAV networks and cloud computing technology the issue of how to acquire big data from sensors is analyzed. On-demand service of cloud-based systems and its impact on UAV control procedures are analyzed. Moreover, stable conditions are derived for UAV cloud control, which defines the relationship between stability of the cloud-based UAV system and data acquisition rate. The results show the effectiveness of the proposed model.

9.5.4 Image Processing

Kim et al. [26] merges the UAV technology to access the cracks in concrete structures. Camera-mounted UAVs are used to provide the images of the structure and the provided information is processed using hybrid image binarization to find the cracks accurately. The proposed system successfully measures cracks thicker than 0.1 mm with a length estimation error of 7.6 at the maximum.

Sankarasrinivasan et al. [27] also analyze crack assessment using UAV technology. In [28] an early fire detection system is designed to monitor the fire events using the concept of sensors network and UAV technology. Wireless sensors technology, UAVs, and cloud computing are the major components of the design to monitor the environmental parameters. Results show that the proposed system has higher fire detection rate than other available methods in literature.

Eisenbeiss and Sauerbier [29] use UAV networks for pacing up the process of information collection in photogrammetry. The authors model a system that uses UAV network for documentation of archaeological excavation. Sharma et al. [28] proposed a fire detection system for smart cities, which is assisted by UAV networks, cloud computing, and image processing techniques. The proposed system has a higher forest fire detection rate of 95–98%.

9.5.5 Millimeter Wave Communication

The authors in [30] integrate mmWave communication with UAV-aided 5G ultra-dense networks, and design a novel link-adaptive constellation-division multiple access (CoDMA) technique. The authors believe that the proposed CoDMA scheme significantly improves the system capacity for ultra-dense 5G networks. UAVs have tremendous potential to improve wireless network capacity, but are challenging to operate in ultra-dense networks, mainly due to the strong interference received from the dominated line-of-sight channels of the UAVs. Owing to the limited number of beams generated in practical mmWave communication systems, conventional beam division multiple access (BDMA) cannot meet the ever-increasing capacity requirement.

In [31], the authors have used a geometric analysis method and greedy user scheduling algorithm in multi-UAV network to detect the blockage. The proposed solution verifies the effectiveness in terms of spectrum efficiency. A sum rate maximization problem is formulated under scheduling, line-of-site, and power constraints. A novel three-dimensional (3D) beam training strategy for UAV-assisted mmWave communications is proposed in [32]. The inverse discrete space Fourier transform is introduced to construct the training beam with flat-topped characteristic. In addition, the hybrid beam forming (BF) system is taken into account and the greedy geometric (GG) algorithm is adopted to obtain the optimal beam. Moreover, compared with other traditional methods, the application of flat-topped beam and GG algorithm effectively improves the accuracy and efficiency of training. The new algorithm can increase the system capacity with low complexity and improve the training efficiency significantly. In the future work, the authors will extend the proposed beam design method from wide-beam scenarios to narrow-beam scenarios. In [33], optimal resource allocation problem has been investigated for downlink coverage from mmWave transmitter mounted on a UAV. In this approach, a circular user space has been divided into multiple sectors motivated by the highly directional beam generated by the antenna array. Side lobe gain of the beams, which causes interference to the other sectors, has been taken into consideration in computing the probability distribution of signal-to-noise-plus-interference ratio. The formulated optimization problem for resource allocation to maximize the sum rate accounts for the power transmission limit, minimum rate guarantee to each user, and backhaul link capacity. Extensive numerical simulations have demonstrated that the proposed algorithm optimally allocates resources to the users, by balancing between interference threshold and minimum rate guarantee. The beam formed by mmWave antenna array is highly directional and requires multiple beam scans to cover the entire area. To limit the interference in concurrent transmission strategy, a threshold on power spillage from adjacent sectors is placed. For this topology, resource allocation problem is formulated aiming to maximize the sum rate while ensuring minimum rate guarantee to each user. It is observed that sum rate variation with height is unimodal.

9.5.6 Artificial Intelligence

In [34] the authors propose a swarm-intelligence-based localization (SIL) and clustering schemes for UAV networks for emergency communications. Particle-optimization-based SIL algorithm is proposed using the bounding box method and then on the basis of this optimization energy-efficient swarm-intelligence-based clustering algorithm is proposed. Results outperform five typical routing protocols related to routing overhead, packet delivery ratio, and end-to-end delay.

Optimization of locations and number of drones for deployment of 5G network in case of emergency and danger (e.g. hurricane disasters, fire accidents, densely populated areas such as stadiums) is carried out in [35]. 5G coverage is optimized while considering the 5G transmitter's coverage range and energy constraints of the drones. Genetic algorithm and simulated annealing algorithms are used to solve the proposed optimization problem and the results are compared. In [36] the authors propose a swarm-intelligence-based autonomous flocking control scheme for UAV network. Distributed Multilayer flocking control scheme, also called swarm intelligence-inspired multi-layer flocking control scheme control scheme (SIMFC), is used to control the flocking using follower and leader drone topology while QoS and low energy consumption are considered. The proposed model is an OMNetCC-based virtual simulator. The results show effective topology control in different scenarios.

9.5.7 Machine Learning

A novel framework is proposed in [37] for integrating reconfigurable intelligent surfaces (RIS) in UAV-enabled wireless networks, where an RIS is deployed for enhancing the service quality of the UAV. Non-orthogonal multiple access (NOMA) technique is invoked to further improve the spectrum efficiency of the network, while mobile users (MUs) are considered as roaming continuously. A decaying deep Q-network (D-DQN)-based algorithm is proposed for tackling this pertinent problem. In the proposed D-DQN-based algorithm, the central controller is selected as an agent for periodically observing the state of UAV-enabled wireless network and for carrying out actions to adapt to the dynamic environment. The energy consumption minimizing problem is formulated by jointly designing the movement of the UAV, phase shifts of the RIS, power allocation policy from the UAV to MUs, as well as determining the dynamic decoding order. In [38], the problem of joint caching and resource allocation is investigated for a network of cache-enabled UAVs that service wireless ground users over the long-term evolution (LTE) licensed and unlicensed bands. The problem is formulated as an optimization problem, which jointly incorporates user association, spectrum allocation, and content caching.

Using the proposed liquid state machine (LSM) algorithm, the cloud can predict the user's content request distribution while having only limited information on the network's and user's states. The results also show that the LSM significantly improves the convergence time of up to 20% compared with conventional learning algorithms such as Q-learning. The problem of joint trajectory design and power control is formulated in [39] for maximizing the instantaneous sum transmit rate while satisfying the rate requirement of users. UAVs have limited payload and flight time; multiple UAVs may have to be harnessed for accomplishing complex high-level tasks, where a control center can be employed for coordinating their actions. Authors in [40] consider image classification tasks in UAV-aided exploration scenarios, where the coordination of multiple UAVs is implemented by a ground fusion center (GFC) positioned in a strategic, but inaccessible location, such as a mountain top, where recharging the battery is uneconomical or may even be infeasible. In order to minimize the computational complexity imposed on the GFC by the UAVs, weighted zero-forcing (WZF) transmit precoding (TPC) is used at each UAV based on realistic imperfect channel state information (CSI). In the proposed FL-aided classification, the local updates of all UAVs are transmitted to the GFC via fading wireless channels. Then, a global update is performed at the GFC and the corresponding results are fed back to UAVs for performing the next round of local model update.

9.5.8 Optimization and Game Theory

Ruan [41] analyzes the coverage problem. The authors propose a multi-UAV coverage model taking into account the energy constraint resulting in energy-efficient communication. Exact potential games that have nash equilibrium are implemented to solve the game theory problem. Spatial adaptive play is adopted to maximize the coverage. The model can be implemented to deploy UAV network in remote or emergency area at optimized locations.

Kusyk et al. [42] use artificial intelligence and game theory to deploy the swarm UAV networks such that they maintain MANET and uniform distribution of the coverage area. Algorithms used for the model are good candidates for response to changes in dynamic environments for tasks such as detection, localization, and tracking. In [43], a game theory approach is taken into account for UAV access selection and BS bandwidth allocation to provide QoS and service cost. For access, allocation among group of UAVs is formulated using dynamic evolutionary game, and bandwidth allocation is done by using non-cooperative game. Stochastic geometry tool is used to model the position distribution of the network nodes. Simulation is done to show the effectiveness of the proposed model. Messous et al. [44] use game theory for computational offloading in UAV networks.

9.6 Research Trends

Presently, some researchers have provided solutions regarding the performance enhancement of the UAV networks for different applications. But still, many unsolved issues should be observed for future studies. In this section, we will observe several research trends for future work. All the issues mentioned have effects on the performance of the UAV network. These are related to network architecture, system integration, and communication links of the system. We are discussing these areas in the context of a disaster environment.

1. *Scalability*: The applicability of UAVs in future applications requires new topologies to be developed. It is very important to observe the scalability factor for mission-critical applications to perform efficiently. For such an application, there may be a need for more UAVs to cover the operational area. In such scenarios, UAV-enabled disaster recovery networks must have the capability to allow more nodes at any time. This requirement of the optimized number of UAV nodes is an important problem to be solved.

2. *UAV platform*: It includes the material selection, position, and orientation of UAVs in the network. For a disturbing environment caused by any natural disaster, some sensitive factors affect the performance of UAVs in the operational area. Owing to the dynamic nature of the environment, sensitivity to weather, temperature, and wind conditions need to be studied. Similarly, the position and orientation of UAV nodes in the network are important key performance parameters.

3. *Privacy*: The distributive nature of UAV nodes in a network brings challenges of privacy. For UAV-based disaster recovery networks, sensors gather information regarding temperature, weather, air, and status of the recovery network. This continuous updating of the operational area is prone to unintentional privacy issues. These concerns about privacy necessitate a special focus on network efficiency. Sending sensitive information to base stations is an open issue to solve.

4. *Data processing*: For a disaster recovery network, constant update of network performance is mandatory, for which there is a need for high data processing. UAVs are equipped with limited storage and computational capability. This high processing of data also requires high power. This power requirement might not be possible for UAV nodes; cloud computing may provide the solution in this regard. Fog and edge computing, being an extension of cloud computing, enhances network performance. The integration of the UAV network with these architectures is an open issue to study.

5. *Routing*: In a disaster recovery network, the environment is constantly changing. There might be more disturbance during the recovery process. This will affect the UAV resource management and topology that was decided at an earlier stage. Routing is one major issue in such a situation. There is a need for routing protocols that keep

the routing table updated constantly, whenever the network topology changes. Path planning: For critical applications such as military and disaster environments, a multi-UAV operation is needed. Such types of applications also require cooperation among UAV nodes to pass the updated information about the status of the operational area. New algorithms are needed for the problem of frequent path changing. The coordination among the UAV nodes also needs to be studied for updated path planning.

6. *QoS provisioning*: UAVs transport different types of data to support many applications. For a disaster recovery network, GPS locations, updating simple text messages, constant video streaming, and voice recording are required. The parameters such as packet loss rate, delay, error, condition of the wireless link, and spectrum allocation affect the network performance. QoS must be maintained for recovery networks for better results. This is one critical area to study, which focuses on fulfilling the QoS requirement.

7. *Energy efficiency*: Drones, swarms, or UAVs are flying machines that can cover all the environments. These devices are light in weight and have limited computational power and energy. To keep the network energy efficient, new algorithms are required. The energy consumption is very high for a disturbing area as compared to other applications. Energy consumption is also related to network lifetime. Optimized solutions are required to cater to the energy minimization problem for UAV-enabled disaster recovery network.

8. *UAV mobility and placement*: Mobility and placement of UAVs in a disturbing operational area have a vast area of research. A disturbing environment requires a multi-UAV operation; these UAV nodes can use different sensors. One might give updates using infrared cameras while another might use another type of camera. This diffusion of data from different types of nodes requires optimized mobility and placement of nodes in the network. Diffusion of data could be done with the surrounding obstacles as well, which damages the actual information. The coordination among all UAVs to collect multiple information at one time is an open research challenge to study.

9. *Handover*: When there is a disaster in an area where there are already many existing obstacles or maybe at the edge of a cellular network or maybe a large area, UAVs must be optimally placed in the network so that such communication could be done efficiently. This problem is related to handover techniques. New handover techniques are required to cater to the problems of long range and obstacles of the operational area.

10. *Integration with existing technologies*: The integration of the UAV network with existing technologies helps in improving the performance of the network. This integration helps many in the completion of many future applications. However, this integration also brings new challenges associated with the operation, scalability, processing, and heterogeneous nature of different devices. This association is an emergent challenge to study what affects the performance of the network to fulfill the application's requirement.

11. *Lack of standards and regulations*: A lot of research needs to be done for defining proper rules, regulations, policies, and guidelines for UAV networks. This will help in the compatibility of UAV networks with the existing technologies.

12. *Security*: The distributive nature of UAVs in a network makes it more secure in such a way that there is no single point of failure. However, a UAV node is itself prone to a variety of cyber threats. Some solutions are required to provide secure communication. For a disaster recovery network, security is an important factor to observe. Strong and robust security solutions are required depending upon the UAV-enabled application. The sensors or programming on UAV nodes are vulnerable to security attacks.

9.7   Future Insights

Based on the benefits of UAV networks, it is foreseen that many sectors will be revolutionized in the future for betterment. This section highlights some future works related to UAV-enabled disaster recovery networks. The following future directions help the academic community to develop optimized networks. We must accelerate research in the following areas.

- To provide a safer and less tense environment for military, and natural disaster recovery applications, sensors on hardware, network architecture, enabling technologies, and processing capability of UAV devices should be represented in new ways.
- Multi-UAV systems should be developed, which maintain the reliability, remote control, and secure transfer of information over the operational area.
- For disaster management, UAV's integration with existing technology must be studied under delay, security, and safety.
- For emergencies such as the epidemiology sector and disaster management, UAVs have the potential to provide solutions regarding low latency and high resolution.

- Until now, there is no proper standardization for UAV networks. Every step from manufacturing, repairing, and marketing of UAV devices must be monitored properly.
- The selection of pilot and owner for a UAV is also an open area for research. Proper registration should be done for better performance for disaster management.
- A UAVs simulator to observe the performance is desired; the simulator will provide real results. For disaster recovery networks, simulation results may not be applicable because of the dynamic emergency environment.
- Investigations are needed to study the issues related to the data collection mechanism. UAV will be deployed over the operational area to collect information and upload to the near cloud or fog server to provide efficient connectivity.
- The mobility factor of multi-UAVs over the area must be observed under the constraints of coverage area and collision avoidance issues.
- Use of other technologies such as artificial intelligence and cloud computing in association with UAVs should be developed to improve the performance in terms of high data rate and low latency.
- For providing the required services over the operational area, The International Telecommunication Union Telecommunication Standardization Sector (ITU-T) should provide standardization for the network development. This standardization will help in the efficient performance of the network for recovery management.

9.8 Conclusion

In this chapter, we have presented a comparative analysis of some existing solutions incorporating UAV technology with other architectures, e.g. cloud computing, machine learning, artificial intelligence, NFV, and SDN. Critical review was done based on QoS factors that affect the performance. This state of the art is followed by sections explaining the design considerations, benefits, and architecture of UAV networks. Table 9.1 can be used to provide solutions for future applications. Most of the work is still waiting to discover the applications of UAV networks.

Bibliography

1 Guha-Sapir, D., Vos, F., Below, R., and Ponserre, S. (2014). Annual Disaster Statistical Review 2011: The Numbers and Trends. *Tech. Rep.*. Centre for Research on the Epidemiology of Disasters (CRED).

2 Reina, D.G., Toral, S., Barrero, F. et al. (2013). Modelling and assessing Ad hoc networks in disaster scenarios. *Journal of Ambient Intelligence and Humanized Computing* 4 (5): 571–579.

3 Reina, D., Askalani, M., Toral, S. et al. (2015). A survey on multihop Ad hoc networks for disaster response scenarios. *International Journal of Distributed Sensor Networks* 11 (10): 647037.

4 García-Campos, J.M., Gutiérrez, D., Sánchez-García, J., and Marn, S.T. (2018). A simulation methodology for conducting unbiased and reliable evaluation of manet communication protocols in disaster scenarios. *Smart Technologies for Emergency Response and Disaster Management*. IGI Global, pp. 106–143.

5 Hayat, S., Yanmaz, E., and Muzaffar, R. (2016). Survey on unmanned aerial vehicle networks for civil applications: a communications viewpoint. *IEEE Communications Surveys & Tutorials* 18 (4): 2624–2661.

6 Sánchez-García, J., García-Campos, J.M., Toral, S. et al. (2016). An intelligent strategy for tactical movements of UAVs in disaster scenarios. *International Journal of Distributed Sensor Networks* 12 (3): 8132812.

7 Van Tilburg, C. (2017). First report of using portable unmanned aircraft systems (drones) for search and rescue. *Wilderness & Environmental Medicine* 28 (2): 116–118.

8 Bekmezci, I., Sahingoz, O.K., and Temel, c.S. (2013). Flying ad-hoc networks (FANETs): a survey. *Ad Hoc Networks* 11 (3): 1254–1270.

9 Marketsandmarkets, unmanned aerial vehicle(uav) market (2018), https://www.marketsandmarkets.com/Market-Reports/unmanned-aerial-vehicles-uav-market-662.html, (Online; accessed 13-12-2020).

10 Acevedo, J.J., Arrue, B.C., Maza, I., and Ollero, A. (2013). Cooperative large area surveillance with a team of aerial mobile robots for long endurance missions. *Journal of Intelligent & Robotic Systems* 70 (1–4): 329–345.

11 De Freitas, E.P., Heimfarth, T., Netto, I.F. et al. (2010). UAV relay network to support WSN connectivity. *International Congress on Ultra Modern Telecommunications and Control Systems*, IEEE, pp. 309–314.

12 Maza, I., Caballero, F., Capitán, J. et al. (2011). Experimental results in multi-UAV coordination for disaster management and civil security applications. *Journal of Intelligent & Robotic Systems* 61 (1–4): 563–585.

13 Semsch, E., Jakob, M., Pavlicek, D., and Pechoucek, M. (2009). Autonomous UAV surveillance in complex urban environments. *2009 IEEE/WIC/ACM International Joint Conference on Web Intelligence and Intelligent Agent Technology*, vol. 2, IEEE, pp. 82–85.

14 George, J., Sujit, P., and Sousa, J.B. (2011). Search strategies for multiple UAV search and destroy missions. *Journal of Intelligent & Robotic Systems* 61 (1–4): 355–367.

15 Hermosilla, A., Zarca, A.M., Bernabe, J.B. et al. (2020). Security orchestration and enforcement in NFV/SDN-aware UAV deployments. *IEEE Access* 8: 131779–131795.

16 Bekkouche, O., Bagaa, M., and Taleb, T. (2019). Toward a UTM-based service orchestration for UAVs in MEC-NFV environment. *2019 IEEE Global Communications Conference (GLOBECOM)*, IEEE, pp. 1–6.

17 Zhao, L., Yang, K., Tan, Z. et al. (2020). A novel cost optimization strategy for SDN-enabled UAV-assisted vehicular computation offloading. *IEEE Transactions on Intelligent Transportation Systems* early access: 1–11.

18 Zhu, Y., Wang, S., Liu, X. et al. (2020). Joint task and resource allocation in SDN-based UAV-assisted cellular networks. *2020 IEEE/CIC International Conference on Communications in China (ICCC)*, IEEE, pp. 430–435.

19 e Silva, T.D., de Melo, C.F.E., Cumino, P. et al. (2019). STFANET: SDN-based topology management for flying ad hoc network. *IEEE Access* 7: 173499–173514.

20 Vishnevsky, V., Kirichek, R., Elagin, V. et al. (2020). SDN-assisted unmanned aerial system for monitoring sensor data. *2020 12th International Congress on Ultra Modern Telecommunications and Control Systems and Workshops (ICUMT)*, IEEE, pp. 313–317.

21 Xiong, F., Li, A., Wang, H., and Tang, L. (2019). An SDN-MQTT based communication system for battlefield UAV swarms. *IEEE Communications Magazine* 57 (8): 41–47.

22 Liu, B., Zhang, W., Chen, W. et al. (2020). Online computation offloading and traffic routing for UAV swarms in edge-cloud computing. *IEEE Transactions on Vehicular Technology* 68 (8): 8777–8791.

23 Khan, U.A., Khalid, W., and Saifullah, S. (2020). Energy efficient resource allocation and computation offloading strategy in a UAV-enabled secure edge-cloud computing system. *2020 IEEE International Conference on Smart Internet of Things (SmartIoT)*, IEEE, pp. 58–63.

24 Chen, W., Liu, B., Huang, H. et al. (2019). When UAV swarm meets edge-cloud computing: the QoS perspective. *IEEE Network* 33 (2): 36–43.

25 Luo, F., Jiang, C., Yu, S. et al. (2017). Stability of cloud-based UAV systems supporting big data acquisition and processing. *IEEE Transactions on Cloud Computing* 7 (3): 866–877.

26 Kim, H., Lee, J., Ahn, E. et al. (2017). Concrete crack identification using a UAV incorporating hybrid image processing. *Sensors* 17 (9): 2052.

27 Sankarasrinivasan, S., Balasubramanian, E., Karthik, K. et al. (2015). Health monitoring of civil structures with integrated UAV and image processing system. *Procedia Computer Science* 54: 508–515.

28 Sharma, A., Singh, P.K., and Kumar, Y. (2020). An integrated fire detection system using IoT and image processing technique for smart cities. *Sustainable Cities and Society* 61: 102332.

29 Eisenbeiss, H. and Sauerbier, M. (2011). Investigation of UAV systems and flight modes for photogrammetric applications. *The Photogrammetric Record* 26 (136): 400–421.

30 Wang, L., Che, Y.L., Long, J. et al. (2019). Multiple access mmwave design for UAV-aided 5G communications. *IEEE Wireless Communications* 26 (1): 64–71.

31 Zhao, J., Liu, J., Jiang, J., and Gao, F. (2020). Efficient deployment with geometric analysis for mmwave UAV communications. *IEEE Wireless Communications Letters* 9 (7): 1115–1119.

32 Zhong, W., Gu, Y., Zhu, Q. et al. (2020). A novel 3D beam training strategy for mmwave UAV communications. *2020 14th European Conference on Antennas and Propagation (EuCAP)*, IEEE, pp. 1–5.

33 Kumar, S., Suman, S., and De, S. (2020). Dynamic resource allocation in UAV-enabled mmWave communication networks. *IEEE Internet of Things Journal* early access: 1.

34 Arafat, M.Y. and Moh, S. (2019). Localization and clustering based on swarm intelligence in UAV networks for emergency communications. *IEEE Internet of Things Journal* 6 (5): 8958–8976.

35 Al-Turjman, F., Lemayian, J.P., Alturjman, S., and Mostarda, L. (2019). Enhanced deployment strategy for the 5G drone-BS using artificial intelligence. *IEEE Access* 7: 75999–76008.

36 Dai, F., Chen, M., Wei, X., and Wang, H. (2019). Swarm intelligence-inspired autonomous flocking control in UAV networks. *IEEE Access* 7: 61786–61796.

37 Liu, X., Liu, Y., and Chen, Y. (2020). Machine learning empowered trajectory and passive beamforming design in UAV-RIS wireless networks. *arXiv preprint arXiv:2010.02749*.

38 Chen, M., Saad, W., and Yin, C. (2019). Liquid state machine learning for resource and cache management in LTE-U unmanned aerial vehicle (UAV) networks. *IEEE Transactions on Wireless Communications* 18 (3): 1504–1517.

39 Liu, X., Liu, Y., Chen, Y., and Hanzo, L. (2019). Trajectory design and power control for multi-UAV assisted wireless networks: a machine learning approach. *IEEE Transactions on Vehicular Technology* 68 (8): 7957–7969.

40 Zhang, H. and Hanzo, L. (2020). Federated learning assisted multi-UAV networks. *IEEE Transactions on Vehicular Technology* 69 (11): 14104–14109.

41 Ruan, L., Wang, J., Chen, J. et al. (2018). Energy-efficient multi-UAV coverage deployment in UAV networks: a game-theoretic framework. *China Communications* 15 (10): 194–209.

42 Kusyk, J., Uyar, M.U., Ma, K. et al. (2020). Artificial intelligence and game theory controlled autonomous UAV swarms. *Evolutionary Intelligence* 1–18, https://doi.org/10.1007/s12065-020-00456-y.

43 Yan, S., Peng, M., and Cao, X. (2018). A game theory approach for joint access selection and resource allocation in UAV assisted IoT communication networks. *IEEE Internet of Things Journal* 6 (2): 1663–1674.

44 Messous, M.-A., Sedjelmaci, H., Houari, N., and Senouci, S.-M. (2017). Computation offloading game for an UAV network in mobile edge computing. *2017 IEEE International Conference on Communications (ICC)*, IEEE, pp. 1–6.

45 Tipantu na, C., Hesselbach, X., Sánchez-Aguero, V. et al. (2019). An NFV-based energy scheduling algorithm for a 5G enabled fleet of programmable unmanned aerial vehicles. *Wireless Communications and Mobile Computing* 2019: 1–20.

46 White, K.J., Denney, E., Knudson, M.D. et al. (2017). A programmable SDN+NFV-based architecture for UAV telemetry monitoring. *2017 14th IEEE Annual Consumer Communications & Networking Conference (CCNC)*, IEEE, pp. 522–527.

47 Nogales, B., Sanchez-Aguero, V., Vidal, I. et al. (2018). A NFV system to support configurable and automated multi-UAV service deployments. *Proceedings of the 4th ACM Workshop on Micro Aerial Vehicle Networks, Systems, and Applications*, pp. 39–44.

48 Niethammer, U., Rothmund, S., Schwaderer, U. et al. (2011). Open source image-processing tools for low-cost UAV-based landslide investigations. *International Archives of the Photogrammetry, Remote Sensing and Spatial Information Sciences* 38 (1): C22.

49 Singhal, C. and Rahul, K. (2019). Efficient QoS provisioning using SDN for end-to-end data delivery in UAV assisted network. *2019 IEEE International Conference on Advanced Networks and Telecommunications Systems (ANTS)*, IEEE, pp. 1–6.

50 Pinto, M.F., Marcato, A.L., Melo, A.G. et al. (2019). A framework for analyzing fog-cloud computing cooperation applied to information processing of UAVs. *Wireless Communications and Mobile Computing* 2019: 1–14.

51 Salhaoui, M., Guerrero-González, A., Arioua, M. et al. (2019). Smart industrial IoT monitoring and control system based on UAV and cloud computing applied to a concrete plant. *Sensors* 19 (15): 3316.

52 Saura, J.R., Reyes-Menendez, A., and Palos-Sanchez, P. (2019). Mapping multispectral digital images using a cloud computing software: applications from UAV images. *Heliyon* 5 (2): e01277.

53 Pehere, S., Sanganwar, P., Pawar, S., and Shinde, A. (2020). Detection of pothole by image processing using UAV. *Journal of Science and Technology* 5 (3): 2456–5660.

54 Cheng, L., Zhu, Q., Wang, C.-X. et al. (2020). Modeling and simulation for UAV air-to-ground mmWave channels. *2020 14th European Conference on Antennas and Propagation (EuCAP)*, IEEE, pp. 1–5.

55 Li, Z., Chen, M., Pan, C. et al. (2019). Joint trajectory and communication design for secure UAV networks. *IEEE Communications Letters* 23 (4): 636–639.

56 Asheralieva, A. and Niyato, D. (2019). Game theory and Lyapunov optimization for cloud-based content delivery networks with device-to-device and UAV-enabled caching. *IEEE Transactions on Vehicular Technology* 68 (10): 10094–10110.

57 Liang, X., Xu, W., Gao, H. et al. (2020). Throughput optimization for cognitive UAV networks: a three-dimensional-location-aware approach. *IEEE Wireless Communications Letters* 9 (7): 948–952.

58 Wang, Y., Li, Z., Chen, Y. et al. (2020). Joint resource allocation and UAV trajectory optimization for space–air–ground internet of remote things networks. *IEEE Systems Journal* early access: 1–11.

59 Fang, S., Chen, G., and Li, Y. (2020). Joint optimization for secure intelligent reflecting surface assisted UAV networks. *IEEE Wireless Communications Letters* 10 (2): 276–280.

60 Feng, W., Tang, J., Zhao, N. et al. (2020). NOMA-based UAV-aided networks for emergency communications. *China Communications* 17 (11): 54–66.

61 Liu, Y., Qin, Z., Cai, Y. et al. (2019). UAV communications based on non-orthogonal multiple access. *IEEE Wireless Communications* 26 (1): 52–57.

62 Yang, Y., Gao, Z., Zhang, Y. et al. (2020). Codeword selection for concurrent transmissions in UAV networks: a machine learning approach. *IEEE Access* 8: 26583–26590.

63 Zhang, Q., Saad, W., Bennis, M. et al. (2020). Predictive deployment of UAV base stations in wireless networks: machine learning meets contract theory. *IEEE Transactions on Wireless Communications* 20 (1): 637–652.

64 Challita, U., Ferdowsi, A., Chen, M., and Saad, W. (2018). Artificial intelligence for wireless connectivity and security of cellular-connected UAVs. *arXiv preprint arXiv:1804.05348.*

65 Liu, X., Chen, M., Liu, Y. et al. (2020). Artificial intelligence aided next-generation networks relying on UAVs. *arXiv preprint arXiv:2001.11958.*

66 Messous, M.-A., Senouci, S.-M., Sedjelmaci, H., and Cherkaoui, S. (2019). A game theory based efficient computation offloading in an UAV network. *IEEE Transactions on Vehicular Technology* 68 (5): 4964–4974.

67 Goudarzi, S., Anisi, M.H., Ciuonzo, D. et al. (2020). Employing unmanned aerial vehicles for improving handoff using cooperative game theory. *IEEE Transactions on Aerospace and Electronic Systems* early access: 1.

10

Network-Assisted Unmanned Aerial Vehicle Communication for Smart Monitoring of Lockdown

Navuday Sharma[1], Muhammad Awais[2], Haris Pervaiz[2], Hassan Malik[3], and Qiang Ni[2]

[1] Test Software Development, Ericsson Eesti AS, Tallinn, Estonia
[2] School of Computing and Communications, Computing and Communications, Lancaster University, Bailrigg, UK
[3] Department of Computer Science, Edge Hill University, Ormskirk, UK

10.1 Introduction

The pandemic COVID-19, commonly known as coronavirus, initiated in Wuhan, China, in December 2019, has become a major concern across the world over the past few months [1]. This respiratory disease is caused by severe acute respiratory syndrome coronavirus-2 (SARS-CoV-2) and affects the victim with breathlessness, fever, cough, and fatigue. According to WHO weekly epidemiological and operational updates, as of 18 October 2020, over 40 million cases and 1.1 million deaths have been reported globally [2]. The spread of coronavirus mainly occurs due to physical contact between an infected (or carrier) and a non-infected person, where there is a possibility of virus transfer to a healthy person's respiratory system. The virus can also be present on different objects in its inactive state when the carrier coughs or sneezes and can enter healthy individuals and become active when they come in contact with that object. Therefore, maintaining social distancing and avoiding physical contact have become the key factors to suppress the spread of COVID-19. To implement such restrictions, Internet-of-things (IoT) architecture can serve as a potential solution, where a variety of wearable and environmental sensors can be deployed to monitor the health and position of individuals and collect data to predict the occurrence of COVID-19 symptoms or contact with a sick individual [3]. However, since COVID-19 has spread globally to a very large extent, the deployment of sensors at such a large scale may not be feasible

Autonomous Airborne Wireless Networks, First Edition.
Edited by Muhammad Ali Imran, Oluwakayode Onireti, Shuja Ansari, and Qammer H. Abbasi.

in a cost-effective and timely manner. Thus, in this chapter, we focus on implementing unmanned aerial vehicles (UAVs)[1] based solutions for monitoring the restrictions of social distancing and physical contact with techniques such as precise localization [4], thermal imaging, and sensor data acquisition. The data obtained can be transferred over the long-term evolution (LTE) or fifth-generation (5G) network to the concerned agencies monitoring the restrictions. The UAVs can perform not only crown surveillance but also mass screening, public announcements, and deliver medical supplies, and reach inaccessible areas.

The integration of UAV network to provide cellular connectivity has gained significant interest in smart city applications [5], particularly for emergency scenarios such as service recovery and disaster relief [6, 7]. The UAVs mounted with transceivers act as aerial base stations (ABSs) or relays connected with the terrestrial cellular network. ABSs can be deployed on demand by the stakeholders [8] to communicate with the smart cameras installed on the street light by hovering over them to provide good-quality link communication for collecting the information about the flash crowds for monitoring the smart lockdown, as shown in Figure 10.1a. The information from the ABSs can be forwarded to the control center via a micro base station (MBS) where the people's identities are matched against the data stored in the government database. A similar approach can be utilized with the help of multi-hop ABS communication to autonomously monitor the traffic during the lockdown as shown in Figure 10.1b. It is important to highlight the importance of the positioning of ABS with regard to the location of smart cameras on the ground for better coverage as depicted clearly by Figure 10.1c, which demonstrates the dependency of ABS communication on the line-of-sight connections. The MBS provides backhaul connectivity to the ABSs and if any ABS is outside the coverage of MBS, it can utilize the neighboring ABSs to reach the MBS, resulting in multi-hop ABS communication.

There are several areas of research for integrating UAVs into 5G and beyond, such as redefined network architecture, developing new physical and network layer techniques such as a study carried out in [5]. In this chapter, we emphasize air-to-ground (A2G) channel measurement and UAV communication in a downlink transmission scheme obtained from ray-tracing results to further enhance the quality of the communication link. In this study, we have investigated two different scenarios such as UAVs acting as ABSs and UAVs acting as relays for terrestrial communication. We implement WinProp suite from Altair and Wireless InSite

---

1 The terms unmanned aerial vehicle (UAV) and aerial base stations (ABSs) are used interchangeably throughout this chapter to improve the readability and clarity purposes.

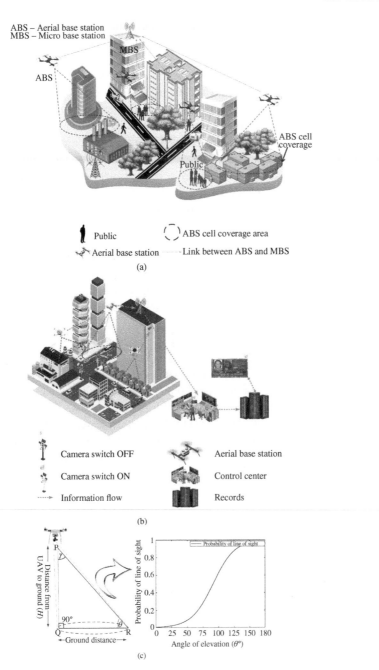

Figure 10.1 Cellular Network-assisted low-altitude aerial base station (ABS) for smart monitoring of lockdown. (a) Autonomous ABSs for smart monitoring of public gatherings. (b) Multi-hop autonomous ABSs for smart traffic monitoring. (c) Line-of-sight dependent ABS communication.

from Remcom as our ray-tracing tools to conduct the radio propagation simulations. Both software are highly reliable when compared with realistic field measurements [9, 10].

The chapter is organized as follows. Section 10.1.1 shows the recent literature addressing the impact of COVID-19 in the current scenario and strategies to find potential solutions with existing communication and computing technologies. Further, we study the two use case scenarios of UAV, as mentioned previously: UAVs as ABSs in Section 10.2 and UAVs as Relays in Section 10.3, and in Sections 10.2.1 and 10.3.2, the simulation setups with ray tracing are described for both scenarios, respectively. Further, Section 10.2.2 outlines the formulation for the optimal number of ABSs to cover a geographical region, given the constraint on ABS transmission power, the altitude of hovering, and including the path loss (PL) and channel fading effects from ray-tracing simulations. The results of such formulation are shown later in Section 10.2.3. Later, for UAV use case as relays, the air interface (AI) of 5G is described in Section 10.3.1, which forms the basis of ray tracing simulation parameters. Further, results of the throughput and power coverage are shown in Section 10.3.2.

10.1.1 Relevant Literature

To combat COVID-19, the telecommunication community has come up with several potential research propositions. Many articles reflect the importance of machine learning and deep learning techniques for diagnosis and prognosis of COVID-19 patients [11–13]. Some research is directed toward the current state, predicting the impact and techniques against COVID-19 in different countries such as Italy [14], Spain [15], South Korea [16], and Bangladesh [17]. Also, many research articles address solutions using smart city technology to reduce the risk of COVID-19. In [18], the authors address the strategies to identify the areas with high human density and mobility using existing cellular network functionalities such as frequency handover and cell selection, thereby impacting social distancing rules to avoid spreading the risk of COVID-19. [19] provides a comprehensive review of artificial intelligence techniques and big data analysis to find COVID-19 diagnosis, treatment, and spreading solutions. Further, beyond-5G-based tactile edge learning framework is addressed in [20], leveraging mobile edge computing, which offers benefits such as low latency, scalability, and privacy. Also, IoT is an essential element of a smart city framework, which has been exploited in [21], where the authors investigate data collection and monitoring for smart pandemic management. Further, associated challenges and evolution in healthcare IoT have also been discussed. In [22], smart technology such as UAVs or robot positioning, non-contact deliveries, and smart healthcare such as face mask recognition, health apps, and telemedicine are discussed for maintaining social distancing between people and avoiding physical contact. Also, to suppress the spreading of coronavirus, social relationships between mobile devices in the social Internet of things (SIoT) is exploited in [23]. Further, a survey on the impact of various technologies such as UAVs, IoT, 5G, and AI has been extensively discussed in [24].

10.2 UAVs as Aerial Base Stations

In this work, we considered UAVs as ABSs as shown in Figure 10.1b, deployed by the operator based on the traffic requirements from cellular network; it can provide cellular coverage to the ground users based on an appropriate A2G channel model [25]. It is important to highlight that the focus is on downlink transmission to provide a good quality communication link to the IoT devices such as smart camera to switch them on and off based on demand to monitor the smart lockdown in the region. A smart camera will be sending the video footage to the ABS, which is then forwarded to the control center at MBS where they can match the individuals using the information stored in the national database. Matching of credentials can be based on the cars or the individuals to highlight the people not following the SOPs of the lockdown as suggested by the government.

However, to obtain the channel characteristics, we implement close-in reference distance path loss model [26], to obtain large-scale fading characteristics such path loss (PL) and shadowing in A2G channel as follows:

$$PL_{LoS}(d)[dB] = 20\log_{10}\left(\frac{4\pi d_0}{\lambda}\right) + 10n_{LoS}\log_{10}(d) + X_{\sigma,LoS}, \quad (10.1)$$

$$PL_{NLoS}(d)[dB] = 20\log_{10}\left(\frac{4\pi d_0}{\lambda}\right) + 10n_{NLoS}\log_{10}(d) + X_{\sigma,NLoS}, \quad (10.2)$$

where, PL is the path loss in dB, d is the radio link distance between the ABS and ground user with $d_0 = 1$ m as the reference distance, λ is the wavelength, n is the path loss exponent, and X_σ is the shadow fading. Also, LoS (in subscript) implicates the parameters when the link is in line-of-sight (LoS) condition and NLoS in non-line-of-sight condition. In particular, the average path loss is given by

$$PL(d)[dB] = \mathbb{P}_{LoS} \cdot PL_{LoS}(d)[dB] + (1 - \mathbb{P}_{LoS}) \cdot PL_{NLoS}(d)[dB].$$

$$(10.3)$$

The variations of large-scale channel parameters such as path loss exponent, shadowing standard deviation, and Rician factor with respect to ABS altitude and transmission power are addressed in [26]. Further, the statistical distributions and spatial correlations of small-scale parameters such as the angle of arrival and angle of departure between the ABS and ground user are shown in [27]. Thus, we can characterize the behavior of A2G downlink channel for the same simulation setup.

10.2.1 Simulation Setting

The simulation was carried out in three different environments – suburban, urban, and urban high rise. These environments were created in 3DS Max according to International Telecommunication Union-Radiocommunication Sector (ITU-R) parameters [28]:

- α = Ratio of land area covered by the buildings of the total area (*dimensionless*).
- β = Mean number of buildings per unit area (*building/km^2*).
- γ = Variable determining the building height distribution.

The simulation area was 1000×1000 m^2, with 32 500 receivers uniformly distributed on the ground and the ABS placed at the center as shown in Figure 10.2. We consider the ABS to be static and only hovering with altitude for our simulations; however, trajectory planning can be done based

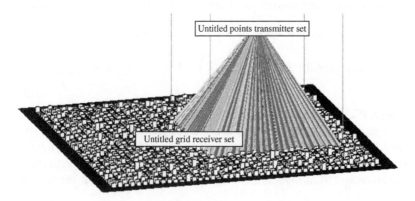

Figure 10.2 Ray tracing simulation in urban environment.

on traffic requirements and signal quality [29]. The density of the buildings depends on the type of environment, with suburban having the least density and urban high rise having the most, while the heights of the buildings were Rayleigh distributed [28]. The altitude of ABS was varied from 100 to 2000 m. Many simulations were carried out at different ABS height intervals of 100 m at 2.4 GHz carrier frequency and 20 MHz signal bandwidth, with a transmission power of ABS ranging from 18 to 46 dBm. Many snapshots of simulations were captured in different parts of the environments. The results of these simulations were averaged at the corresponding heights and transmission power to increase the accuracy of the results. The simulations performed in Wireless InSite were considered to be accurate in comparison with practically measured results [10].

10.2.2 Optimal Number of ABSs for Cellular Coverage in a Geographical Area

The number of ABSs for providing cellular coverage to a geographical area depends on the altitude, transmission power of the ABS, and type of propagation environment. When the altitude of the ABS is low, less transmission power is needed to cover an area due to lower propagation attenuation. However, when the altitude is high, more transmission power is needed to cover the same area. This implies that there should be a trade-off between the altitude and the transmission power of the ABS. Therefore, an optimal altitude and transmission power are required to obtain the best coverage for a particular propagation environment. These optimal values can be seen from [30] and its mathematical formulation in [31]. Here, we identify the number of ABSs at different heights and transmission power of ABS for suburban, urban, and urban high-rise environments. For this, we first calculate the coverage by ABS using (10.4). We find the ABS coverage as a portion of the geographical area over which the power received by the mobile devices is above the threshold.

$$C = \frac{1}{A_c} \int_{A_C} r \cdot P(P_{rx}(r) \geq P_{\min}) \, dr \, d\phi, \tag{10.4}$$

where r is the distance between each receiver and ABS, ϕ is the azimuth angle, P is the probability that the received power P_{rx} is greater than the threshold $P_{\min}$, and A_C is the assumed cell area for normalizing the final coverage area. From the coverage results of ABS from (10.4), the number of ABSs needed can simply be found from (10.5).

$$\text{Number of ABS} = \frac{\text{Total Geographical area to be covered}}{\text{Coverage area by ABS}}. \tag{10.5}$$

From here, we find the variation in the number of the ABSs with ABS altitude and transmitting power for a fixed geographical area. For variation with respect to geographical area, simulations were performed for a fixed altitude and transmitting power.

10.2.3 Performance Evaluation

In this section, we show the results for a number of ABSs required to provide cellular coverage to a geographical area at different ABS heights and transmitting power in suburban, urban, and urban high-rise environments. These results are based on the coverage analysis by ABS as discussed in Section 10.2.2. Here, we show the comparative analysis for receiver thresholds kept at −120 and −100 dBm. For better coverage performance with higher receiver thresholds, similar trend variations were seen as between −120 and −100 dBm, mainly for suburban environments due to different fading scenarios than urban environments, as explained in the later sections. Also, a variation in the number of ABSs with an increase in a geographical area with fixed ABS height and transmission power is shown.

10.2.3.1 Variation of Number of ABSs with ABS Altitude

Figure 10.3 shows the variation of number of ABSs with changing ABS altitude. These results are based on keeping the geographical area with a radius of 200 m. However, it was observed that increasing the geographical area does not change the behavior of the curves, but only the number of ABSs required. From (10.5), we can see that the number of ABSs required will only depend on the coverage by each ABS since the geographical area to be covered is assumed constant here. At −120 dBm threshold, i.e. minimum received signal strength indicator (RSSI) for LTE mobile device, it was observed that the number of ABSs needed at lower altitude is more due to less coverage since few receivers could only satisfy the probability condition given in (10.4) because of multipath and scattering propagation effects. When the altitude is increased, the number of ABSs needed reduces to a certain extent where an optimal altitude of the ABS is obtained for maximum coverage, i.e. 350–400 m [30] and minimum number. At higher altitudes, the number of ABSs again starts increasing since the path loss effect leads to higher signal attenuation, thereby reducing the coverage from each ABS. Also, the number of ABSs needed for the suburban environment is lower than urban and urban high-rise environments. This is because Rayleigh fading scenario was observed from ray-tracing simulations and Zajić [32] in suburban environments, leading to increased coverage area

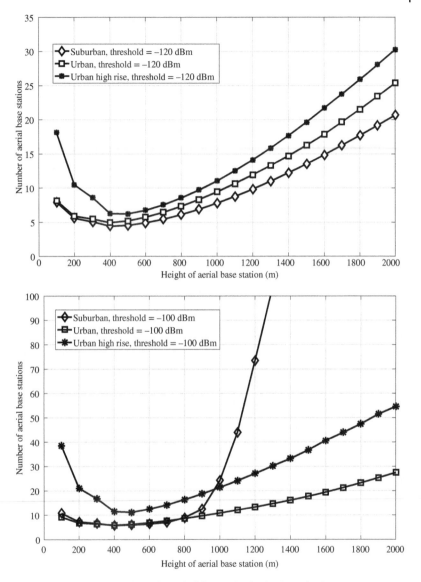

Figure 10.3 Variation of number of ABSs required with its altitude.

whereas Rician fading was seen for urban environments, for A2G channel. Compared with a higher threshold, −100 dBm, similar behavior was observed for urban and urban high-rise environments, with higher number of ABSs than lower threshold result. This is due to the higher capacity and better connectivity required by mobile users, which results in more ABSs.

Table 10.1 Fitting parameters for receiver threshold −120 dBm.

Parameter	Suburban	Urban	Urban high rise
a	9.88E−06	8.548E−07	2.157E−06
b	−8.151	−9.637	−9.761
c	8.202	9.81	11.66
d	0.5984	0.6124	0.6202

Table 10.2 Fitting parameters for receiver threshold −100 dBm.

Parameter	Suburban	Urban	Urban high rise
a	8.556E−09	4.297E−07	1.08E−06
b	18.66	−10.1	−10.68
c	—	11.15	22.08
d	—	0.5755	0.5995

However, in a suburban environment, an exponential increment in the number of ABSs was observed, since due to Rayleigh fading and higher path loss at high altitudes, the received power by mobile devices further reduces. The variation of the number of ABSs, Ω with its altitude, h, can be represented as an exponential function.

$$\Omega = a \times \exp(b \times h) + c \times \exp(d \times h), \tag{10.6}$$

where a, b, c, and d are fitting parameters as given on Tables 10.1 and 10.2.

10.2.3.2 Variation of Number of ABS with ABS Transmission Power

The analysis for this variation can be seen in Figure 10.4. This analysis was done at an optimal ABS altitude of 320 m for maximum ABS coverage and minimum number of ABSs needed. Theoretically, the increase in transmission power should increase the ABS cell coverage area, thereby reducing the number of ABSs needed. However, the number of ABSs remains mostly constant in the simulation runs of the suburban environment at −120 dBm and slightly reduces when the threshold was increased to −100 dBm. At higher thresholds, more reduction in the number of ABSs needed was observed. This is because the receivers with −120 dBm threshold will be at the cell boundary and already receiving the least possible power needed to maintain connectivity with the ABS. Owing to the Rayleigh fading, high signal strength attenuation was observed because of multipath

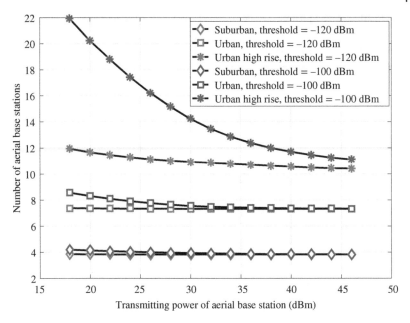

Figure 10.4 Variation of number of ABSs required with its transmitting power.

and scattering effects. Therefore, increasing the transmission power does not increase cellular coverage as expected. But at −100 dBm threshold, the receivers can afford the reduction of received power, without losing connectivity with the ABS. Therefore, more reduction in the number of ABSs was seen at higher thresholds with increasing transmission power. This remains valid for the urban and urban high-rise environments as well. However, the rate of reduction can be seen more in an urban high rise, then urban, and least in suburban. This is again due to Rician fading, in urban environments that enable the line-of-sight components to reach a farther distance with higher transmission power, with more ABS coverage. As in Section 10.2.3.1, we also provide a similar analytical expression for the number of ABS, Ω with ABS transmitting power, P.

$$\Omega = a' \times \exp(b' \times P) + c' \times \exp(d' \times P), \tag{10.7}$$

where a', b', c' and d' are fitting parameters as given on Tables 10.3 and 10.4.

10.2.3.3 Variation of Number of ABSs with Geographical Area

In Figure 10.5, the variation of the number of ABS with the geographical area to be covered is shown. These simulations were performed at a fixed altitude and transmission power of ABS. However, similar behavior for the curves

Table 10.3 Fitting parameters for receiver threshold −120 dBm.

Parameter	Suburban	Urban	Urban high rise
a'	0.000394	0.002649	0.01987
b'	−2.647	−1.911	−2.237
c'	3.825	7.331	10.82
d'	−0.0002713	−5.503E−05	−0.02689

Table 10.4 Fitting parameters for receiver threshold −100 dBm.

Parameter	Suburban	Urban	Urban high rise
a'	0.4818	0.3378	8.613
b'	−0.4185	−0.9818	−0.4974
c'	3.42	7.14	4.789
d'	0.0292	0.01231	0.2619

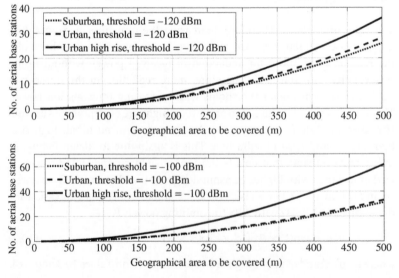

Figure 10.5 Variation of number of ABSs required with geographical area to be covered.

Table 10.5 Fitting parameters for receiver threshold −120 dBm.

Parameter	Suburban	Urban	Urban high rise
a''	386.3	−7592	−1579
b''	0.3457	0.3387	0.3372
c''	−379.4	7599	1588
d''	0.3332	0.3394	0.3414

Table 10.6 Fitting parameters for receiver threshold −100 dBm.

Parameter	Suburban	Urban	Urban high rise
a''	1.028E+05	−851.3	268.4
b''	0.3228	0.3354	0.361
c''	−1.028E+05	860	−252
d''	0.3227	0.3426	0.3169

was obtained at other altitudes and transmission power. As is expected, the number of ABSs needed to cover a given geographical area increases with the increase in a geographical area, provided that the ABS coverage remains constant. Also, as discussed previously suburban environments will need less ABSs due to Rayleigh fading than urban environments in Rician fading. Analytically, this curve also follows an exponential behavior as (10.6) and (10.7), with geographical area A.

$$\Omega = a'' \times \exp(b'' \times A) + c'' \times \exp(d'' \times A), \tag{10.8}$$

where a'', b'', c'' and d'' are fitting parameters as given in Tables 10.5 and 10.6.

10.3 UAV as Relays for Terrestrial Communication

Consider a terrestrial base station (TBS) with 5G antenna that communicates with the master UAV. It is a dynamic directional antenna that follows the position of the UAV to provide data to the modem of the transceiver installed on the UAV. The TBS antenna is placed 3m above the ground. The prediction of the whole simulation is taken at a height of 150m cruising altitude of the ABS (Figure 10.6).

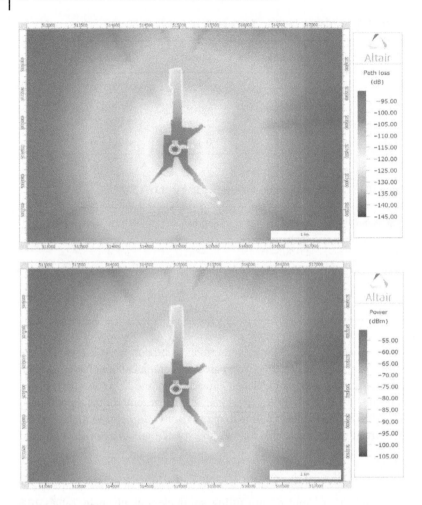

Figure 10.6 TBS path loss and transmission power.

The master UAV has been used as a modem/router whose signal strength and power have been predicted in the ray-tracing simulation. The master UAV guides the slave unmanned aerial vehicles (SUAVs) cluster, set at a height of 100 m, which moves accordingly with it to reduce and theoretically eliminate the Doppler shift caused by the movement. Owing to the high altitude of the master UAV, the power of the signal is equally located over the whole predicted area (Figure 10.7), which is formally the cruising height of the SUAVs 100 m. The status of LoS link between master UAV and SUAVs is also shown in Figure 10.7.

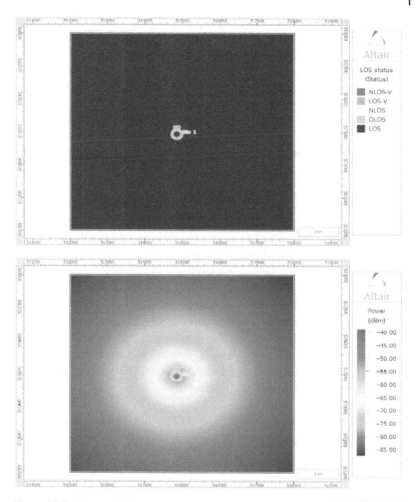

Figure 10.7 Master UAV transmission power and line of sight.

The SUAVs are used to provide Internet communication to the users on the ground level (1.5 m). For this system, simulation has been done for six UAVs whose combined power of the antennas can sufficiently guarantee a full channel performance over a zone of 4 km^2 as shown in Figure 10.9.

10.3.1 5G Air Interface

The air interface is the access mode that characterizes the communication link between two different stations in a mobile or wireless communication. The air interface involves both the physical and data link layers. The physical connection of an AI is generally radio based. This is a link between the base station (BS) and a mobile station (MS). Multiple links can be created in a limited spectrum through different multiple access techniques such as frequency division multiple access (FDMA), time division multiple access (TDMA), or space division multiple access (SDMA). Some advanced forms of transmission multiplexing combine frequency and time division approach such as orthogonal frequency division multiple access (OFDM), or code division multiple access (CDMA).

There is a multitude of 5G applications [33], broadly classified under three domains: enhanced mobile broadband (EMBB), in which large volumes of data are required by the users with high data rates; massive machine type communication (MMTC), where a small amount of data is needed for longer duration; and ultra reliable low latency communication (URLLC) which demands data transfer with very low round trip latency but without any data loss. Therefore, implementing a common air interface for multi-service scenarios is not a feasible solution and possibly flexible configured orthogonal frequency division multiple access (FC-OFDM) can serve as a suitable contender, which can flexibly configure different sub-band characters [34]. The commercial use of the 5G standard is supposed to be deployed in 2020. For every new mobile generation, higher frequency bands are assigned and wider spectral bandwidth per frequency channel to increase throughput, i.e. 30 kHz for 1G, 200 kHz for 2G, 5 MHz for 3G, 20 MHz for 4G, and 100 MHz for 5G.

10.3.2 Simulation Setup

For every simulation, Altair WinProp suite, specifically ProMan and WallMan, has been used in order to perform an Intelligent Ray Tracing (IRT) prediction. The ProMan (Propagation) software package is designed to predict the PL accurately between transmitter and receiver including all important parameters of the mobile radio channel. The ProMan software offers network planning modules for 2G/2.5G, 3G/3.5, 4G/LTE, WLAN, and WiMAX communication protocols. Static network planning modules, as well as dynamic network simulators, are included. Besides the cellular network planning features, ProMan also supports the planning of broadcasting networks (terrestrial and satellite). The three further simulations aim to represent and collect data of a real system deployment based on the use of UAVs. The system is divided into three different channel blocks as shown in Figure 10.8. For the simulations, the parameters used have been shown in Table 10.7.

Figure 10.8 System for channel measurement.

Table 10.7 5G air interface simulation parameters.

Multiple access	OFDMA
Maximum number of sub-carriers	2048
Guard sub-carriers	255
Sub-carrier spacing	75 kHz
Symbol duration	13.33 µs
Sub-carriers on resource block	14
Channel bandwidth	100 MHz
Carrier frequency	28 GHz
MIMO	Up four streams
Duplex separation	FDD
Resource block for QAM	100

Milan, Italy scenario has been used as an urban environment with high building density with a PL coefficient $\eta = 5$. For all the simulations the antenna is an isotropic radiator with a transmission power of 40 dBm, a transmitter antenna gain of 25 dB, an azimuth equal to zero and a down-tilt of zero, and vertical polarization. The transmission power and PL of TBS and master UAV are shown in Figures 10.6 and 10.7, respectively. One of the most important components to design and analyze the link budget of a communication system is the large-scale fading effects such as PL and shadowing. The network simulation shows from Table 10.8 that with these assumptions the required performance is satisfied, where 500 Mbps of data

Table 10.8 Download maximum throughput.

BPSK	65.63 Mbps
16-QAM	262.5 Mbps
32-QAM	382.1 Mbps
64-QAM	393.8 Mbps
128-QAM	459.4 Mbps

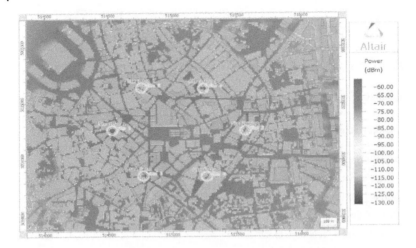

Figure 10.9 Received power by ground users from SUAVs cluster.

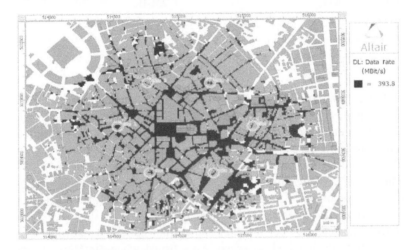

Figure 10.10 64-QAM throughput coverage area.

rate is achieved using the 128-QAM modulation scheme. Further, received power and throughput by the ground user from SUAVs transmission can be observed in Figures 10.9 and 10.10, respectively.

10.4 Conclusion

In this chapter, we discuss the importance of UAVs to monitor COVID-19 restrictions of social distancing, public gatherings, and physical contacts in a smart city environment. We address two use case scenarios where UAVs are deployed by the telecom operators as ABSs and relays for terrestrial communication networks. We implemented both scenarios with a ray-tracing simulation setup with two radio propagation software: WinProp Suite and Wireless InSite. Later, a detailed discussion on channel measurement formulation and 5G air interface is presented. Further, to evaluate the key performance indicators such as cell coverage, throughput, and received power of ground user for both scenarios detailed simulation results are discussed. The simulation result shows the trade-off between altitude and transmission power of the ABS with respect to the number of ABSs required to cover a particular geographical area. It is shown that with the increase in altitude of ABS, the number of ABSs required to cover a geographical area increases proportionally.

On the other hand, with the increase in transmission power at the ABS, the number of ABSs required to cover the same geographical area decreases proportionally. However, the rate of reduction can be seen more in an urban high rise, then urban and least in suburban. This is due to the Rician fading in urban environments, which enable the line-of-sight components to reach a farther distance with higher transmission power, with more ABS coverage. Moreover, it was observed that increasing the geographical area does not change the behavior of the curves, but only the number of ABSs required. As a whole, it can be concluded that with the optimal selection of the number of ABSs, transmission power, and altitude, ABS can be one of the potential solutions to cover the geographical area for crowd surveillance and monitoring.

Bibliography

1 WHO (2020). Coronavirus disease (Covid-19) pandemic. https://www
.who.int/emergencies/diseases/novel-coronavirus-2019 (accessed 09
March 2021).

2 WHO (2020). Covid-19 weekly epidemiological update. https://www.who
.int/docs/default-source/coronaviruse/situation-reports/20201020-weekly-
epi-update-10.pdf (accessed 09 March 2021).

3 Vedaei, S.S., Fotovvat, A., Mohebbian, M.R. et al. (2020). COVID-SAFE:
an IoT-based system for automated health monitoring and surveillance
in post-pandemic life. *IEEE Access* 8: 188538–188551. https://doi.org/10
.1109/ACCESS.2020.3030194.

4 Arafat, M.Y. and Moh, S. (2019). Localization and clustering based on
swarm intelligence in UAV networks for emergency communications.

IEEE Internet of Things Journal 6 (5): 8958–8976. https://doi.org/10.1109/JIOT.2019.2925567.

5 Li, B., Fei, Z., and Zhang, Y. (2019). UAV communications for 5G and beyond: recent advances and future trends. *IEEE Internet of Things Journal* 6 (2): 2241–2263.

6 Masood, A., Scazzoli, D., Sharma, N. et al. (2020). Surveying pervasive public safety communication technologies in the context of terrorist attacks. *Physical Communication* 41: 101109. https://doi.org/https://doi.org/10.1016/j.phycom.2020.101109.

7 Masood, A., Sharma, N., Alam, M.M. et al. (2019). Device-to-device discovery and localization assisted by UAVs in pervasive public safety networks. *Proceedings of the ACM MobiHoc Workshop on Innovative Aerial Communication Solutions for FIrst REsponders Network in Emergency Scenarios*, pp. 6–11.

8 Sharma, N., Magarini, M., Jayakody, D.N.K. et al. (2018). On-demand ultra-dense cloud drone networks: opportunities, challenges and benefits. *IEEE Communications Magazine* 56 (8): 85–91. https://doi.org/10.1109/MCOM.2018.1701001.

9 Hoppe, R., Wölfle, G., Futter, P., and Soler, J. (2017). Wave propagation models for 5G radio coverage and channel analysis. *2017 Sixth Asia-Pacific Conference on Antennas and Propagation (APCAP)*, pp. 1–3. https://doi.org/10.1109/APCAP.2017.8420499.

10 Mede?ovi?, P., Veleti?, M., and Blagojevi?, . (2012). Wireless insite software verification via analysis and comparison of simulation and measurement results. *2012 Proceedings of the 35th International Convention MIPRO*, May 2012, pp. 776–781.

11 Hu, S., Gao, Y., Niu, Z. et al. (2020). Weakly supervised deep learning for Covid-19 infection detection and classification from CT images. *IEEE Access* 8: 118869–118883. https://doi.org/10.1109/ACCESS.2020.3005510.

12 Wang, X., Deng, X., Fu, Q. et al. (2020). A weakly-supervised framework for Covid-19 classification and lesion localization from chest CT. *IEEE Transactions on Medical Imaging* 39 (8): 2615–2625. https://doi.org/10.1109/TMI.2020.2995965.

13 Oh, Y., Park, S., and Ye, J.C. (2020). Deep learning Covid-19 features on CXR using limited training data sets. *IEEE Transactions on Medical Imaging* 39 (8): 2688–2700. https://doi.org/10.1109/TMI.2020.2993291.

14 Tropea, M. and De Rango, F. (2020). Covid-19 in Italy: current state, impact and ICT-based solutions. *IET Smart Cities* 2 (2): 74–81. https://doi.org/10.1049/iet-smc.2020.0052.

15 Cecilia, J.M., Cano, J., Hernández-Orallo, E. et al. (2020). Mobile crowdsensing approaches to address the Covid-19 pandemic in Spain. *IET Smart Cities* 2 (2): 58–63. https://doi.org/10.1049/iet-smc.2020.0037.

16 Ahn, N.Y., Park, J.E., Lee, D.H., and Hong, P.C. (2020). Balancing personal privacy and public safety during Covid-19: the case of South Korea. *IEEE Access* 8: 171325–171333. https://doi.org/10.1109/ACCESS.2020.3025971.

17 Islam, M.N. and Islam, A.K.M.N. (2020). A systematic review of the digital interventions for fighting Covid-19: the Bangladesh perspective. *IEEE Access* 8: 114078–114087. https://doi.org/10.1109/ACCESS.2020.3002445.

18 Alsaeedy, A.A.R. and Chong, E.K.P. (2020). Detecting regions at risk for spreading Covid-19 using existing cellular wireless network functionalities. *IEEE Open Journal of Engineering in Medicine and Biology* 1: 187–189. https://doi.org/10.1109/OJEMB.2020.3002447.

19 Pham, Q., Nguyen, D.C., Huynh-The, T. et al. (2020). Artificial intelligence (AI) and big data for coronavirus (Covid-19) pandemic: a survey on the state-of-the-arts. *IEEE Access* 8: 130820–130839. https://doi.org/10.1109/ACCESS.2020.3009328.

20 Rahman, M.A., Hossain, M.S., Alrajeh, N.A., and Guizani, N. (2020). B5G and explainable deep learning assisted healthcare vertical at the edge: Covid-19 perspective. *IEEE Network* 34 (4): 98–105. https://doi.org/10.1109/MNET.011.2000353.

21 Ndiaye, M., Oyewobi, S.S., Abu-Mahfouz, A.M. et al. (2020). IoT in the wake of Covid-19: a survey on contributions, challenges and evolution. *IEEE Access* 8: 186821–186839. https://doi.org/10.1109/ACCESS.2020.3030090.

22 Jaiswal, R., Agarwal, A., and Negi, R. (2020). Smart solution for reducing the Covid-19 risk using smart city technology. *IET Smart Cities* 2 (2): 82–88. https://doi.org/10.1049/iet-smc.2020.0043.

23 Wang, B., Sun, Y., Duong, T.Q. et al. (2020). Risk-aware identification of highly suspected Covid-19 cases in social IoT: a joint graph theory and reinforcement learning approach. *IEEE Access* 8: 115655–115661. https://doi.org/10.1109/ACCESS.2020.3003750.

24 Chamola, V., Hassija, V., Gupta, V., and Guizani, M. (2020). A comprehensive review of the Covid-19 pandemic and the role of IoT, drones, AI, blockchain, and 5G in managing its impact. *IEEE Access* 8: 90225–90265. https://doi.org/10.1109/ACCESS.2020.2992341.

25 Khawaja, W., Guvenc, I., Matolak, D.W. et al. (2019). A survey of air-to-ground propagation channel modeling for unmanned aerial vehicles. *IEEE Communications Surveys Tutorials* 21 (3): 2361–2391. https://doi.org/10.1109/COMST.2019.2915069.

26 Sharma, N., Magarini, M., Dossi, L. et al. (2018). A study of channel model parameters for aerial base stations at 2.4 GHz in different environments. *2018 15th IEEE Annual Consumer Communications Networking Conference (CCNC)*, pp. 1–6. https://doi.org/10.1109/CCNC.2018.8319165.

27 Sharma, N., Magarini, M., Reggiani, L., and Alam, M.M. (2019). Channel characterization at 2.4 GHz for aerial base station. *Procedia Computer Science* 151: 1092–1099. https://doi.org/https://doi.org/10.1016/j.procs.2019.04.155. *The 10th International Conference on Ambient Systems, Networks and Technologies (ANT 2019) / The 2nd International Conference on Emerging Data and Industry 4.0 (EDI40 2019) / Affiliated Workshops.*

28 ITU-R (2003). Propagation data and prediction methods for the design of terrestrial broadband millimetric radio access systems. Geneva, Switzerland, Rec. P.1410-1412, P Series, Radiowave Propagation.

29 Sheikh, M.U., Riaz, M., Jameel, F. et al. (2020). Quality-aware trajectory planning of cellular connected UAVs. *Proceedings of the 2nd ACM MobiCom Workshop on Drone Assisted Wireless Communications for 5G and Beyond*, pp. 79–85.

30 Cileo, D.G., Sharma, N., and Magarini, M. (2017). Coverage, capacity and interference analysis for an aerial base station in different environments. *2017 International Symposium on Wireless Communication Systems (ISWCS)*, Aug 2017, pp. 281–286. https://doi.org/10.1109/ISWCS.2017.8108125.

31 Sharma, N., Sharma, V., Magarini, M. et al. (2019). Cell coverage analysis of a low altitude aerial base station in wind perturbations. *2019 IEEE Globecom Workshops (GC Wkshps)*, pp. 1–6. https://doi.org/10.1109/GCWkshps45667.2019.9024665.

32 Zajić, A. (2012). *Mobile-to-Mobile Wireless Channels*. Artech House.

33 Navarro-Ortiz, J., Romero-Diaz, P., Sendra, S. et al. (2020). A survey on 5G usage scenarios and traffic models. *IEEE Communications Surveys Tutorials* 22 (2): 905–929. https://doi.org/10.1109/COMST.2020.2971781.

34 Lin, H. (2015). Flexible configured OFDM for 5G air interface. *IEEE Access* 3: 1861–1870. https://doi.org/10.1109/ACCESS.2015.2480749.

11

Unmanned Aerial Vehicles for Agriculture: an Overview of IoT-Based Scenarios

Bacco Manlio[1], Barsocchi Paolo[1], Gotta Alberto[1], and Ruggeri Massimiliano[2]

[1]*National Research Council (CNR), Institute of Information Science and Technologies (ISTI) and Institute of Science and Technologies for Energy and Sustainable Mobility, Pisa, Italy*
[2]*National Research Council (CNR), Institute of Science and Technologies for Energy and Sustainable Mobility, Ferrara, Italy*

11.1 Introduction

Smart farming (SF) refers to the application of Information and Communication Technology (ICT) to agriculture. Data collected and analyzed through ICT techniques support efficient production processes, thus motivating scientists, practitioners, and private and public companies to work toward the goal of developing and encouraging the use of innovative technologies to support farmers on the ground. The most relevant technologies and techniques to be fully exploited are the satellite imagery, the use of agricultural robots, a larger use of sensor nodes to collect data, and the potentialities of unmanned aerial vehicles UAVs for remote sensing and actuation [1]. Remote sensing, especially UAV-based, is used in a plethora of different scenarios, as for instance forestry [2]. Choosing the sensors to be installed on remote sensing systems depends on the scenario. Satellites represent a long-standing solution for remote sensing, and IoT scenarios can also be found applied to the satellite field [3]. Contrarily to satellites, anyway, UAVs provides great flexibility from the point of view of payloads, because those can be changed at every flight. UAVs must be considered as a breakthrough technology, a game changer [4, 5]. The potential is tremendous, especially when combined with IoT and low-power wide area networks (LPWANs). A categorization of the scenarios enabled by the combined use of these paradigms sees the use of data acquisition and data analysis to feed decision support systems (DSSs), or to enable semi-autonomous solutions, or even

Autonomous Airborne Wireless Networks, First Edition.
Edited by Muhammad Ali Imran, Oluwakayode Onireti, Shuja Ansari, and Qammer H. Abbasi.

semi-autonomous solutions. The most common types of IoT sensors used in agriculture can gather data on water in the soil, moisture content, electrical conductivity, and acidity; along with those, solutions well established in other fields are also used, such as wind, temperature, humidity, and solar radiation sensors. A case of special interest is the detection of weeds through optical components combined with machine vision systems [6].

UAVs in agriculture can provide support on various issues - for instance, in reconstructing 3D models of crops, determining the crop height, or estimating the value of agricultural indexes, such as the leaf area index (LAI) or the normalized difference vegetation index (NDVI), taking advantage of, e.g. multispectral cameras. Very precise imagery can be collected, and large fields can be monitored through the use of swarms [7, 8] that can be used as flying ad hoc networks (FANETs) [9]. This helps in the detection of diseases, yield estimation, pest monitoring, and creation of virtual plantations, for instance. It is also worth citing phenotyping, and UAVs have shown potential in this scenario, because the collected imagery can be used by image analysis techniques to improve the throughput.

The aim of this work is to present and discuss both research initiatives and scientific literature on the topic of IoT-based SF, especially looking at the use of UAVs in this field. The rest of this work is structured as follows: In Section 11.2, we analyze how UAVs are used in research projects and what the application scenarios considered are. In Section 11.3, the analysis on the application scenarios is deepened, taking into account also selected scientific works from the literature, highlighting the role of unmanned vehicles. In Section 11.4, both requirements and solutions for networking are presented, briefly comparing existing protocols supporting IoT scenarios in agricultural settings. Then, Section 11.5 opens to the potential future role of the joint use of multi-access edge computing (MEC) and 5G networks, presenting network architectures to connect smart farms through satellites and UAVs. Finally, the conclusions can be read in Section 11.6.

11.2 The Perspective of Research Projects

In this section, we will survey the relevant research projects recently funded in the European Union (EU) in the field of SF, especially those exploiting UAVs. The main goal is to highlight the increasing attention toward those activities and to analyze the technologies involved, beyond describing the application scenarios of interest.

In the last years, the EU has been actively undertaking R&I activities, laying the ground for the digitization of European agriculture, also taking into account the potential role that UAVs can play. Strategic interventions have been funded to support the uptake of digital technologies and to develop new solutions. In Table 11.1, we show the most recent EU projects that exploit UAVs, detailing the project name and its starting date, the objective to be achieved from the point of view of the application scenario, and the digital paradigms exploited. The aim is to highlight the projects explicitly taking into account the use of UAVs for different goals. As expected, most projects rely on the use of multiple techniques and technologies in an integrated manner; herein, we highlight the most prominent ones.

The use of unmanned vehicles is a trend of great interest, as also confirmed by looking at the projects in Table 11.1. In the following, we describe more extensively each initiative. The *APMAV* project consists of an intuitive solution for agricultural management based on UAV technology and an intelligent cloud-based platform that provides farmers valuable, actionable, and real-time recommendations for driving down costs and improving crop performance. The *Flourish* project leverages on UAVs as well, aiming at surveying a field from the air, and then at performing a targeted intervention on the ground with an unmanned ground vehicle (UGV). The idea is to develop a DSS targeting precision farming (PF) application with minimal user intervention. The *SWAMP* project develops IoT-based methods and approaches for smart water management in the precision irrigation domain, in order to utilize water more efficiently and effectively, avoiding both under- and over-irrigation.

Data-driven activities are proposed in the *Dragon* project, whose main efforts are directed toward skill transfers to ease the adoption of PF techniques. Several data sources are considered and the data flows are analyzed through the use of Big Data techniques to provide agricultural knowledge and to develop information systems. This project leverages on the joint use of several techniques in an ambitious manner. The *PANTHEON* project, by taking advantage of the technological advancements in the fields of robotics, remote sensing, and Big Data management, aims at designing an integrated system where a limited number of heterogeneous unmanned robotic components (including terrestrial and aerial robots) move within the orchards to collect data and perform some of the most common farming operations. The *SWEEPER* project has proposed a robotic system to harvest sweet peppers in greenhouses, leveraging on machine vision techniques to acquire both color and distance information, and then storing the collected items in an onboard container. Another robotic platform has been developed in the *ROMI* project in an open and lightweight fashion. Assisting in weed reduction and crop monitoring, these robots reduce manual labor and increase the productivity. Land robots also acquire detailed information on sample plants and are

Table 11.1 The most relevant EU-funded R&I projects exploiting UAV technology in the farming sector in the last years.

Project	Start date ended (yes/no)	Goal(s)	Cloud/edge computing	Data serv. inform. sys.	Sensing Terr.	Sat.	Unmanned v. Aerial	Terr.	Data analysis Big data	ML
Sweeper	February 2015 (y)	Harvest robot			X		X	X		
Flourish	March 2015 (y)	Crop monit.	X		X		X	X		X
RUC-APS	October 2020 (n)	Agric. prod.	X				X			
IoF2020	January 2017 (y)	crop Monit. Livestock farm. Dairy monit.	X	X	X		X	X	X	
APMAV	March 2017 (y)	Crop monit.	X		X		X		X	X
Romi	November 2017 (n)	Crop monit.					X	X		X
Pantheon	November 2017 (n)	Orchard monit. Water use			X		X	X	X	
Swamp	November 2017 (y)	Water use	X	X	X		X		X	
BigDataGrapes	January 2018 (y)	Crop monit.	X	X	X		X		X	X
Dragon	October 2018 (n)	Crop monit. Skill acquisition	X	X	X	X	X		X	X

coupled with a UAV that acquires complementary information at the crop level. The *BigDataGrapes* project focuses on grapes and wine production, using machine learning (ML) to support decisions. Data are collected also through UAVs, as for instance canopy characteristics. The *RUC-APS* project is centered on management approaches aiming at enhancing SF solutions in agriculture systems, applying operational research to optimize farm production. The *IoF2020* project is a very large initiative from the point of view of SF digital technologies: in particular, this project aims at accelerating the adoption of IoT, in order to secure sufficient, safe, and healthy food and at strengthening competitiveness of farming and food chains in Europe. It aims at fostering a symbiotic ecosystem of farmers, food industry, technology providers, and research institutes. As evident from Table 11.1, unmanned vehicles can be used to cover different scenarios, such as monitoring, harvesting, and supporting decisions in uncertain conditions. Their use is typically coupled with the use of sensors on the ground, and with cloud solutions to collect, store, and analyze the data to extract actionable information. In Section 11.3, IoT scenarios in the agricultural field are presented, highlighting the role that UAVs can play.

11.3 IoT Scenarios in Agriculture

The use of UAVs in rural areas is less constrained by regulations than in urban areas [10], and thus their use is increasing at a faster rate [11]. This is mainly due to the lower density of both people and buildings, which also translates into fewer obstacles for the vehicle itself. On the other hand, the degree of connectivity is lower, forcing to prefer the use of dedicated solutions instead of relying on existing networks, such as the cellular ones. Distance from airports is likely greater than in urban areas, thus posing less constraints on flights, which can be performed in a complete preprogrammed way, requiring minimal intervention from the pilot. Anyway, the presence of the pilot is a requirement almost everywhere in the world, limiting the scope of beyond visual line of sight (BVLoS) flights. The utility of UAVs, as well as other technologies, lies in transforming agricultural practices, giving birth to smart farms. In what follows, we concentrate on the scientific literature concerning the use of UAVs in several agricultural scenarios, especially with IoT-based setups. The results are presented in Tables 11.2 and 11.3, the former providing details on the considered works, and the latter detailing the agricultural scenarios of interest.

Higher mechanization in agricultural processes has been fueled by the large diffusion of low-cost sensor nodes and the increasing use of actuating

Table 11.2 Surveyed literature in the field of SF, especially considering the use of unmanned vehicles in IoT scenarios.

Work(s)	Main objective(s)
[6]	Survey on machine vision on board autonomous agr. vehicles
[7]	Autonomous inspections for PF operations
[8]	Farm Management Information System (FMIS) for PF
[12]	Survey on UAVs applications for agr.
[13]	Remote sensing and image analysis for agr. UAVs
[14]	UAVs to determine management zones for soil sampling purposes
[15]	Application scenarios and known limitations of agr. UAVs
[16]	Survey on IoT use in agr. scenarios
[17]	UAV to estimate plowing depth with an RGB-D sensor
[18]	UAV to distinguish sugar beets from weeds
[19]	UAV and terrestrial sensing to measure leaf temperature
[20]	UAV for precision spraying
[21]	802.15.4 channel modeling for UAVs use in fields
[22]	UAV in viticulture to find missing plants
[23, 24]	Aerial and terrestrial robots in agr. and forestry
[25]	Survey on commercial UAVs platforms for SF
[26]	Long-range networking for IoT scenarios in agriculture
[27]	IoT greenhouse management system
[28, 29]	IoT irrigation system with long-range networking
[30]	IoT monitoring system in agr.
[31]	Yield and response to fertilizers with UAVs

solutions. Real-time stream processing, analysis, and reasoning are key concepts toward automation in this field [23], i.e. toward a larger use of robots that can adapt to space and time-varying conditions with minimal delay. Robots can perform very precise operations, and can operate in fleets, as proposed in [24], which considers both UGVs and UAVs. Moving systems rely on Global Navigation Satellite System (GNSS) techniques for precise positioning, and PF applications need large accuracy: because of this, the real-time kinematic (RTK) technique is used to improve location accuracy. Several commercial systems integrate a GNSS receiver and use one or more fixed RTK reference base stations [32] for providing accuracy up to centimeters. Beyond precise positioning, robots depend on machine vision systems

Table 11.3 Agricultural scenarios covered by the described works and the use of unmanned vehicles.

Scenario	Work(s)	Use of unmanned vehicles	
		Aerial	*Terrestrial*
Plowing evaluation	[17]	✓	
Spraying	[20, 24]	✓	✓
Monitoring	[7, 8, 13, 15, 16, 22]	✓	
Soil/field mapping and analysis	[14, 15, 23]	✓	✓
Seeding/planting	[15]	✓	
Weeding	[6, 18, 23]	✓	✓
Irrigation/water management	[15]	✓	
Health assessment	[12, 13]	✓	
Yield estimation	[31]	✓	

to navigate the environment [6]; according to the technology and the scenario under consideration, specific spectral signatures are of interest (for instance, (normalized difference vegetation index [NDVI]), and hyperspectral imagery is today a reality for both local and remote sensing. Commercial devices, to be used on board, can already capture both RGB and near infrared (NIR) bands, and stereovision systems can be used to build 3D maps of the environments [6].

UAVs can be used as data mules to collect data from nodes in open fields [21] or to assess whether an area has been subjected to plowing and the plowing depths. In [17], the authors' consider the use of UAVs as an alternative to satellites. In fact, according to the authors, even high-resolution satellites cannot classify the roughness of the terrain, thus motivating the use of UAVs. An RGB camera is installed on board and the collected georeferenced data are analyzed to assess the plowing depths. RGB and NIR are collected by means of a UAV also in [18], with the aim of classifying plants and weeds. The proposed system makes use of the excess green index (ExG) [18] in the case of RGB-only, which depends on the green, red, and blue color components in the images; if NIR is exploited as well, NDVI can be used because of the richer information it provides. By combining these results with geometric features, sugar beets can be recognized even in the case of overlapping plants; furthermore, weed detection is reported as accurate in the space among the rows of sugar beets. NDVI has been used also in the segment of viticulture for precision applications [22]: in fact, when a UAV platform is used to collect very detailed images in a vineyard, plant rows can be discriminated from inter-rows, thus being able to identify missing plants with very good precision.

UAVs can be seen as part of a wireless sensor network (WSN), acting as mobile nodes [19]; thus being able to analytically characterize the channel model between a moving UAV and fixed terrestrial nodes becomes of interest [21]. In particular, PF takes advantage of UAVs (several commercial systems are already available [25]) ranging from fixed to rotary wing machines, able to fly at different speeds and altitudes. Several uses are already in place for UAVs, such as the aforementioned monitoring scenarios, beyond pesticide spraying, which is a key application for PF [20]. Heavy and large UAVs can be used for such a purpose in the case of large fields, combined with the use of multispectral techniques in order to firstly generate an NDVI map, and then using this information to efficiently spray pesticides and fertilizers. The idea of a more accurate use of resources, avoiding any waste and using the right amounts, has been around for quite some time.

11.3.1 Use of Data and Data Ownership

As highlighted in Section 11.2, data are at the very core of the transformation of the agricultural process. Nowadays, the right to access and use the collected data is at the center of the discussion: Europe has recently made steps toward granting the data originator (i.e. the farmer) a leading role in controlling the access and the use of data by means of the so-called *Code of Conduct on Agricultural Data Sharing* by COPA-COGECA.

11.4 Wireless Communication Protocols

Wireless data transfers from both sensors and vehicles in smart agriculture scenarios play a key role, as mostly application scenarios are based on the use of WSNs. This section provides an overview on the most adopted wireless technologies in this field.

Regarding agriculture, short- and long-range communication standards can play different roles: they may be used in an integrated manner in complex scenarios, or separately according to the specific application. Communication standards can be classified according to the achievable datarate instead of the coverage range; the datarate may be crucial for bandwidth-hungry applications. Analogously, data rate has a relative impact on energy requirements, that is, high data rates such as those used for imaging and videos, as well as large bulk of data, entail high energy consumption; differently, in the case of long-time monitoring and small volumes of data, wireless protocols with lower data rates assure a long-lasting operation both in the case of long- and short-range communications.

Considering the large set of wireless communication protocols and the necessities in smart agriculture applications, the use of LPWAN protocols is widespread. Such a protocol family is designed to provide affordable connectivity on large coverage areas and with a relatively low energy profile. Long-range wide area network (LoRaWAN), Sigfox, and NarrowBand Internet of things (NB-IoT) are the leading technologies nowadays [33]. LoRaWAN has been created by the LoRa Alliance and takes advantage of the LoRa modulation with Chirp Spread Spectrum modulation [34], which is easy and affordable to implement on LoRa chips. LoRa transceivers operate at sub-GHz frequencies (e.g. 868 MHz in Europe, 915 MHz in USA). LoRaWAN uses a star topology to interconnect the nodes, and to collect sensing data through a central *concentrator*. It converts these data in order to be transferred through the Internet. It is specifically designed for IoT applications to connect thousands of sensors, modules, and appliances over a large network. LoRaWAN is commonly utilized in a variety of agriculture application areas including weather forecasting [35], irrigation control [28], and farm monitoring [26]. Sigfox is an ultra-narrowband wireless protocol for low data rate applications, thereby making this technology appropriate for IoT and machine-to-machine (M2M) systems [36]. Sigfox implements the Differential Binary Phase Shift Keying (D-BPSK) modulation and operates in unlicensed ISM bands, i.e. 868 MHz in Europe, 915 MHz in North America, and 433 MHz in Asia. It is an LPWAN network protocol that offers an end-to-end IoT connectivity solution based on its patented technologies. Sigfox is also suitable for agricultural applications, such as monitoring silo and tank levels, measuring the temperature of grain stocks, protecting remote farmhouses and outbuildings, securing gates and deterring livestock thieves, optimizing colony health with remotely monitored beehives, and monitoring food temperatures along the entire cold chain.[1] Finally, NB-IoT is considered as the latest radio access technology that emerges from 3rd Generation Partnership Project (3GPP) to enable support for IoT devices. Differently from LoRaWAN and Sigfox, NB-IoT coexists with Global System for Mobile (GSM) and Long-Term Evolution (LTE) under licensed frequency bands 900 MHz [37] with a frequency bandwidth of 200 kHz [38]. Thanks to long battery life, large coverage, and its low cost, NB-IoT is a very suitable solution for various agriculture applications such as livestock tracking, greenhouse monitoring [39], and precision

---

[1] Scenarios at: https://www.sigfox.com/en/agriculture.

farming [40]. For the sake of completeness, other technological solutions for the smart agriculture – aside from such dedicated protocols – have been implemented on top on very diffused wireless protocols, i.e. Wi-Fi and Zigbee.

Wi-Fi is the most commonly utilized protocol in many indoor and outdoor applications: it is considered as the main option for various IoT applications for smart agriculture such as data collection and the connection to cloud. Nevertheless, the high power consumption level of Wi-Fi makes it poorly suitable for agriculture scenarios, despite being still largely adopted. Wi-Fi operates at the 2.4 or 5 GHz frequency bands. Its transmission range is almost limited, up to 1 km, but it provides a nominal minimum bit rate of 1 Mb/s at 2.4 GHz and can scale up for a large set of bandwidth-demanding applications. ZigBee is a wireless personal area network (WPAN) communication technology that was specifically developed by ZigBee Alliance based on the 802.15.4 for low cost and low power solutions for home automation [41]. Nevertheless, thanks to the low cost and the great diffusion of

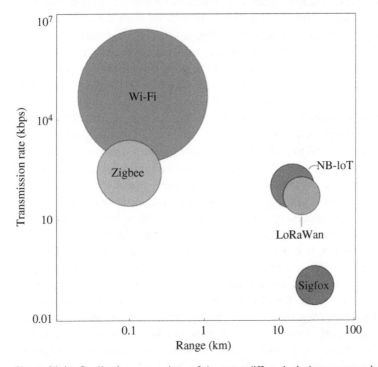

Figure 11.1 Qualitative comparison of the most diffused wireless communication protocols in smart agriculture presented in terms of transmission rate and range (axes) and power consumption (bubble size).

commercial off-the-shelf (COTS) chips, Zigbee has been firmly imposed in various applications such as greenhouse monitoring [21, 27], water saving [29], and yield improvement [30].

There are other features of wireless network protocols to be considered because they can be more significant than others in terms of requirements. Indeed, some smart agriculture applications might need adequate data transfer rates and large coverage ranges due to high numbers of sensors spread over the field. Besides, it is typically unpractical to provide fixed power to the sensors, thus justifying the use of batteries. Anyway, frequent replacement of batteries would not be desirable. Hence, power consumption is another key parameter for sensors, and the power used for wireless transmissions must be carefully weighted. Considering these key features, Figure 11.1 provides a qualitative comparison among the cited network protocols in terms of range and transmission rate (axes), and of power consumption (bubble size).

11.5 Multi-access Edge Computing and 5G Networks

The scope of this section is to discuss the potential advantages brought by the exploitation of the MEC paradigm in the SF context.

The cloud computing paradigm enables the offloading of data and data processing to remote servers. Such an approach, intrinsically centralized, is showing its limits with respect to the needs of today's services and applications. In particular, many IoT applications require mobility management, location awareness, low latency, and scalability; those requirements cannot be fully satisfied by centralized approaches. Several challenges should be taken into account, such as single point of failure, lack of location awareness, loss of reachability, and latency, which can severely impact the expected performance level [42]. Furthermore, centralized servers can be overwhelmed by huge amounts of traffic, thus adding further delays and suffering efficiency loss: as a matter of fact, the very rapid growth of cloud-based applications and services is putting the central infrastructure under pressure [42]. Because of those reasons, edge computing has been proposed as the candidate for filling the gap, meaning that intermediate entities are placed among the end-devices and the cloud. On-site MEC servers process the data collected from close sensors, thus removing the need for upload to a remote cloud service. By doing so, the overall delay is reduced, and reliable services are provided with high availability. IoT, which is considered one of the key use cases of MEC [43], is expected to

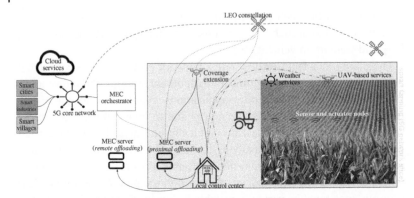

Figure 11.2 Plausible network architectures, highlighting the use of MEC in *proximal offloading* and *remote offloading* configurations [46].

greatly benefit from it. In some ways, MEC, standardized by European Telecommunications Standards Institute (ETSI), can be seen as a new era of the cloud, where a decentralized infrastructure opens new markets and enables real-time critical applications, such as vehicle to anything (V2X) scenarios. When coupled with the features offered by 5G, such as the network slicing [44] highlighted in Figure 11.2, and the increasing softwarization process, MEC and software defined networking (SDN) together bring efficient network operation and service delivery [45]. The popularity of SDN is speeding up the adoption of MEC, and we propose in what follows a high-level network architecture of interest in the context of SF.

When looking at rural contexts, where Internet connectivity still suffers from poor coverage in several geographic areas, aerospace networks have played a role in filling the gap [9] since a long time. The advent of mega-low earth orbit (LEO) constellations and nano-satellites, as complementary to geosynchronous (GEO) backbones [46], is expected to fuel the massive introduction of digital technologies in the SF context.[2] Along with connectivity via satellites, MEC-based architectures exploiting satellites have been proposed in [46, 47], in order to take advantage of MEC even in contexts with poor terrestrial coverage, or to deliver multimedia flows with the desired quality of experience (QoE). In Figure 11.2, which depicts a plausible scenario, two possible reference architectures are considered, according to the scenario under consideration. In the case of small-scale agriculture, a local MEC server could be economically unfeasible; thus remote offloading [46] represents the best option: while the satellite delay must still be paid for,

---

2 See the case of the Starlink constellation of LEO satellites by SpaceX.

relying on LEO satellites greatly reduces such a value (one-way delays can be lowered to less than 50 ms, well below the average 270 ms in the case of GEO satellites), and the presence of edge servers close to the terrestrial satellite gateway (or to the radio network controller site [48]) shrinks to the minimum both the terrestrial network and the processing delays. In the case of large-scale agriculture, where larger investments are expected because of the potential larger revenue, proximal offloading [46] becomes economically feasible; such a solution offers very low delay and computational power close to where it is needed.

In [49], the authors propose UAVs with onboard MEC functionalities, taking into account the constraint due to limited batteries capacity. The tasks to be performed are split into two parts: a fraction to be run on board the UAV, and the other fraction to be migrated toward terrestrial stations. MEC is exploiting the advent of 5G, and designing those functionalities on board UAVs benefits large commercial interests, but needs careful design analysis. Mobile-enabled and wireless infrastructure aerial drones are considered in [50], i.e. drones exploiting 5G connectivity in the former case and drones relaying or providing 5G connectivity in the latter one. The authors are convinced that, although integrating drones into cellular networks is a rather complicated issue, 5G standard has progressed in building the fundamental support mechanisms.

Referring to Figure 11.2, we assume that 5G connectivity is offered via LEO satellites, and that a MEC server acts as an intermediate entity for satisfying the aforementioned requirements of a real-time IoT network, for instance relying on autonomous vehicles and operations, such as coverage extension and agricultural services. Connectivity can be provided in two different manners: through a direct connection to satellites, which means that the latter have onboard transparent interfaces; or through the use of relay nodes. In Figure 11.2, both options are considered: the case of a direct connection of the control center to the satellite transparent payload [51], and the case of a UAV, acting as a relay node for coverage extension [52]. At this time, the use of relay nodes is the most doable solution for both technical and economic reasons.

Looking more closely at the use of IoT in the example proposed in Figure 11.2, both short- and long-range sensor nodes are deployed in the field, as for instance those exploiting the long range (LoRa) protocol, which cover long distances even in difficult conditions. The data collected by sensors are then delivered toward a server for analysis and storing purposes. Several heterogeneous data sources, i.e. sensors placed in the soil and measuring, e.g. the use of water, satellite/UAV imagery, and weather-related data coming from stations placed in the field, are collected altogether. Being heterogeneous sources, both in the data type and in the generation time-scale, there is the need to integrate and analyze raw data at a software level, with the clear purpose of deriving useful information to be used by a farmer or by autonomous vehicles. In the former case, such a system acts as a DSS for the farmer's needs; in the latter case, such a system acts as an (un)supervised centralized intelligence, guiding the autonomous vehicles during their operations or sending commands to the deployed actuators. The UAVs and UGVs can be preloaded with the needed information before departing, as common nowadays, or can exchange data, to be used by the central intelligence as real-time feedback while guiding the vehicle. Given the requirements of the aforementioned autonomous case, the availability of a MEC server is crucial to achieving very low delays, and to efficiently elaborating large quantities of raw data, as in the case of large-scale agriculture. Thus, UAVs can efficiently support environmental monitoring [53] and agriculture applications [4, 21]. As already discussed, civil applications require that the human pilot, still needed for security reasons, has visual line of sight (VLoS) with the UAV, thus somewhat limiting the scope of their use because of the reduced operating range. The possibility of using cellular networks to extend the range to use UAVs through 4G/5G links has been the subject of studies in the literature [54, 55], in order to characterize the quality of the signal received at flying altitudes [55]. In [56], the authors test the possibility of using LTE networks for piloting an UAV in a rural area in Denmark. The preliminary measurement campaigns, aimed at characterizing the path loss model, are then backed by simulation results confirming that LTE networks can be used for such a purpose. In [57], the authors proposed the use of a UAV-to-ground video feed to support a remote pilot on the ground through visual context while piloting in BVLoS conditions. The video feed and the telemetry are supposed to be complementary and mixed together in the data flow coming from the UAV to the ground control station (GCS). Such a study is framed in the wider context of using swarms of drones for different purposes, like either monitoring of power lines, or of cultural heritage sites, or for agricultural applications.

11.6 Conclusion

Agriculture is experiencing a new revolution, but the potential for such a phenomenon to succeed depends on several factors. In this work, we focus on the technological factors, such as IoT; unmanned systems for data gathering, coverage extension, and for agriculture-specific application scenarios; long- and short-range networking solutions for data collection; 4G/5G cellular networks and satellite solutions for connectivity; and on the potential of MEC in the field. This short survey is aimed at providing a holistic view of all the technologies turning around such a digital revolution in Agriculture 4.0.

Bibliography

1 Bacco, M., Barsocchi, P., Ferro, E. et al. (2019). The digitisation of agriculture: a survey of research activities on smart farming. *Array* 3: 100009.

2 Torresan, C., Berton, A., Carotenuto, F. et al. (2017). Forestry applications of UAVs in Europe: a review. *International Journal of Remote Sensing* 38 (8–10): 2427–2447.

3 Bacco, M., Boero, L., Cassara, P. et al. (2019). IoT applications and services in space information networks. *IEEE Wireless Communications* 26 (2): 31–37.

4 Bacco, M., Berton, A., Ferro, E. et al. (2018). Smart farming: opportunities, challenges and technology enablers. *IoT Vertical and Topical Summit on Agriculture-Tuscany (IOT Tuscany)*, IEEE, pp. 1–6.

5 Bacco, M., Brunori, G., Ferrari, A. et al. (2020). IoT as a digital game changer in rural areas: the DESIRA conceptual approach. *2020 Global Internet of Things Summit (GIoTS)*, IEEE, pp. 1–6.

6 Pajares, G., García-Santillán, I., Campos, Y. et al. (2016). Machine-vision systems selection for agricultural vehicles: a guide. *Journal of Imaging* 2 (4): 34.

7 Doering, D., Benenmann, A., Lerm, R. et al. (2014). Design and optimization of a heterogeneous platform for multiple UAV use in precision agriculture applications. *IFAC Proceedings* 47 (3): 12272–12277.

8 Zhai, Z., Martínez Ortega, J.-F., Lucas Martínez, N., and Rodríguez-Molina, J. (2018). A mission planning approach for precision farming systems based on multi-objective optimization. *Sensors* 18 (6): 1795.

9 Bacco, M., Cassará, P., Colucci, M. et al. (2017). A survey on network architectures and applications for nanosat and UAV swarms. *International Conference on Wireless and Satellite Systems (WISATS)*, EAI, pp. 1–10.

10 Reger, M., Bauerdick, J., and Bernhardt, H. (2018). Drones in agriculture: current and future legal status in Germany, the EU, the USA and Japan. *Landtechnik* 73 (3): 62–79.

11 Patel, P. (2016). Agriculture drones are finally cleared for takeoff. *IEEE Spectrum* 53 (11): 13–14.

12 Radoglou-Grammatikis, P., Sarigiannidis, P., Lagkas, T., and Moscholios, I. (2020). A compilation of UAV applications for precision agriculture. *Computer Networks* 172: 107148.

13 Kulbacki, M., Segen, J., Knieć, W. et al. (2018). Survey of drones for agriculture automation from planting to harvest. *2018 IEEE 22nd International Conference on Intelligent Engineering Systems (INES)*, IEEE, pp. 000353–000358.

14 Huuskonen, J. and Oksanen, T. (2018). Soil sampling with drones and augmented reality in precision agriculture. *Computers and electronics in agriculture* 154: 25–35.

15 Yinka-Banjo, C. and Ajayi, O. (2019). Sky-farmers: applications of unmanned aerial vehicles (UAV) in agriculture. In: *Autonomous Vehicles*, (ed: G. Dekoulis). IntechOpen.

16 Boursianis, A.D., Papadopoulou, M.S., Diamantoulakis, P. et al. (2020). Internet of things (IoT) and agricultural unmanned aerial vehicles (UAVs) in smart farming: a comprehensive review. *Internet of Things* 100187.

17 Tripicchio, P., Satler, M., Dabisias, G. et al. (2015). Towards smart farming and sustainable agriculture with drones. *International Conference on Intelligent Environments (IE)*, IEEE, pp. 140–143.

18 Lottes, P., Khanna, R., Pfeifer, J. et al. (2017). UAV-based crop and weed classification for smart farming. *International Conference on Robotics and Automation (ICRA)*, IEEE, pp. 3024–3031.

19 Moribe, T., Okada, H., Kobayashl, K., and Katayama, M. (2018). Combination of a wireless sensor network and drone using infrared thermometers for smart agriculture. *15th Annual Consumer Communications & Networking Conference (CCNC)*, IEEE, pp. 1–2.

20 Mogili, U.R. and Deepak, B. (2018). Review on application of drone systems in precision agriculture. *Procedia Computer Science* 133: 502–509.

21 Bacco, M., Berton, A., Gotta, A., and Caviglione, L. (2018). IEEE 802.15.4 air-ground UAV communications in smart farming scenarios. *IEEE Communications Letters* 22 (9): 1910–1913.

22 Matese, A. and Di Gennaro, S.F. (2018) Practical applications of a multisensor UAV platform based on multispectral, thermal and RGB high resolution images in precision viticulture. *Agriculture* 8 (7): 1–13.

23 Roldán, J.J., del Cerro, J., Garzón-Ramos, D. et al. (2017). Robots in agriculture: state of art and practical experiences. In: *Service Robots*, (ed: A. J. R. Neves). IntechOpen.

24 Gonzalez-de Santos, P., Ribeiro, A., Fernandez-Quintanilla, C. et al. (2017). Fleets of robots for environmentally-safe pest control in agriculture. *Precision Agriculture* 18 (4): 574–614.

25 Puri, V., Nayyar, A., and Raja, L. (2017). Agriculture drones: a modern breakthrough in precision agriculture. *Journal of Statistics and Management Systems* 20 (4): 507–518.

26 Citoni, B., Fioranelli, F., Imran, M.A., and Abbasi, Q.H. (2019). Internet of things and LoRaWAN-enabled future smart farming. *IEEE Internet of Things Magazine* 2 (4): 14–19.

27 Tafa, Z., Ramadani, F., and Cakolli, B. (2018). The design of a ZigBee-based greenhouse monitoring system. *2018 7th Mediterranean Conference on Embedded Computing (MECO)*, IEEE, pp. 1–4.

28 Fraga-Lamas, P., Celaya-Echarri, M., Azpilicueta, L. et al. (2020). Design and empirical validation of a LoRaWAN IoT smart irrigation system. *Multidisciplinary Digital Publishing Institute Proceedings*, Volume 42, p. 62.

29 Ma, W., Wei, Y., Sun, F., and Li, Y. (2019). Design and implementation of water saving irrigation system based on Zigbee sensor network. In: *IOP Conference Series: Earth and Environmental Science*, vol. 252, 052086. IOP Publishing, http://dx.doi.org/10.1088/1755-1315/252/5/052086.

30 Zhou, Z., Xu, K., and Wu, D. (2016). Design of agricultural internet of things monitoring system based on ZigBee. *Chemical Engineering Transactions* 51: 433–438.

31 Schut, A.G., Traore, P.C.S., Blaes, X., and Rolf, A. (2018). Assessing yield and fertilizer response in heterogeneous smallholder fields with UAVs and satellites. *Field Crops Research* 221: 98–107.

32 Thomasson, J.A., Baillie, C.P., Antille, D.L. et al. (2019). Autonomous technologies in agricultural equipment: a review of the state of the art. *American Society of Agricultural and Biological Engineers*. Distinguished lecture no. 40, 1–17, Louisville Kentucky USA. ASABE publication number 913c0119.

33 Mekki, K., Bajic, E., Chaxel, F., and Meyer, F. (2019). A comparative study of LPWAN technologies for large-scale IoT deployment. *ICT Express* 5 (1): 1–7.

34 Bankov, D., Khorov, E., and Lyakhov, A. (2016). On the limits of LoRaWAN channel access. *2016 International Conference on Engineering and Telecommunication (EnT)*, IEEE, pp. 10–14.

35 Bacco, M., Barsocchi, P., Cassará, P. et al. (2020). Monitoring ancient buildings: real deployment of an IoT system enhanced by UAVs and virtual reality. *IEEE Access* 8: 50131–50148.

36 Lavric, A., Petrariu, A.I., and Popa, V. (2019). Long range SigFox communication protocol scalability analysis under large-scale, high-density conditions. *IEEE Access* 7: 35816–35825.

37 Adhikary, A., Lin, X., and Wang, Y.-P.E. (2016). Performance evaluation of NB-IoT coverage. *2016 IEEE 84th Vehicular Technology Conference (VTC-Fall)*, IEEE, pp. 1–5.

38 Wang, Y.-P.E., Lin, X., Adhikary, A. et al. (2017). A primer on 3GPP narrowband Internet of Things. *IEEE communications magazine* 55 (3): 117–123.

39 He, C., Shen, M., Liu, L. et al. (2018). Design and realization of a greenhouse temperature intelligent control system based on NB-IoT. *Journal of South China Agricultural University* 39 (2): 117–124.

40 Castellanos, G., Deruyck, M., Martens, L., and Joseph, W. (2020). System assessment of WUSN using NB-IoT UAV-aided networks in potato crops. *IEEE Access* 8: 56823–56836.

41 Baronti, P., Pillai, P., Chook, V.W. et al. (2007). Wireless sensor networks: a survey on the state of the art and the 802.15.4 and ZigBee standards. *Computer Communications* 30 (7): 1655–1695.

42 Porambage, P., Okwuibe, J., Liyanage, M. et al. (2018). Survey on multi-access edge computing for internet of things realization. *IEEE Communications Surveys & Tutorials* 20 (4): 2961–2991.

43 Sabella, D., Vaillant, A., Kuure, P. et al. (2016). Mobile-edge computing architecture: the role of MEC in the Internet of Things. *IEEE Consumer Electronics Magazine* 5 (4): 84–91.

44 Konstantinos, S., Costa-Perez, X., and Sciancalepore, V. (2016). From network sharing to multi-tenancy: the 5G network slice broker. *IEEE Communications Magazine* 54 (7): 32–39.

45 Blanco, B., Fajardo, J., Giannoulakis, I. et al. (2017). Technology pillars in the architecture of future 5G mobile networks: NFV, MEC and SDN. *Computer Standards & Interfaces* 54 (4): 216–228.

46 Zhang, Z., Zhang, W., and Tseng, F.-H. (2018). Satellite mobile edge computing: improving QoS of high-speed satellite-terrestrial networks using edge computing techniques. *IEEE Network* 33 (1): 70–76.

47 Ge, C., Wang, N., Selinis, I. et al. (2019) QoE-assured live streaming via satellite backhaul in 5G networks. *IEEE Transactions on Broadcasting*. 65 (2): 381–391.

48 Patel, M., Naughton, B., Chan, C. et al. (2014). Mobile-edge computing: introductory technical white paper. *MEC Industry Initiative* 1089–7801, https://www.etsi.org/images/files/etsiwhitepapers/etsi_wp11_mec_a_key_technology_towards_5g.pdf.

49 Hua, M., Huang, Y., Sun, Y. et al. (2019). Energy optimization for cellular-connected UAV mobile edge computing systems. *International Conference on Communication Systems (ICCS)*, IEEE, pp. 1–6.

50 Bor-Yaliniz, I., Salem, M., Senerath, G., and Yanikomeroglu, H. (2019). Is 5G ready for drones: a look into contemporary and prospective wireless networks from a standardization perspective. *IEEE Wireless Communications* 26 (1): 18–27.

51 Guidotti, A., Vanelli-Coralli, A., Conti, M. et al. (2019) Architectures and key technical challenges for 5G systems incorporating satellites. *IEEE Transactions on Vehicular Technology* 68 (3): 2624–2639.

52 Amorosi, L., Chiaraviglio, L., D'Andreagiovanni, F., and Blefari-Melazzi, N. (2018). Energy-efficient mission planning of UAVs for 5G coverage in rural zones. *International Conference on Environmental Engineering (EE)*, IEEE, pp. 1–9.

53 Bacco, M., Delmastro, F., Ferro, E., and Gotta, A. (2017). Environmental monitoring for smart cities. *IEEE Sensors Journal* 17 (23): 7767–7774.

54 Van der Bergh, B., Chiumento, A., and Pollin, S. (2016). LTE in the sky: trading off propagation benefits with interference costs for aerial nodes. *IEEE Communications Magazine* 54 (5): 44–50.

55 Lin, X., Yajnanarayana, V., Muruganathan, S.D. et al. (2018). The sky is not the limit: LTE for unmanned aerial vehicles. *IEEE Communications Magazine* 56 (4): 204–210.

56 Nguyen, H.C., Amorim, R., Wigard, J. et al. (2017). Using LTE networks for UAV command and control link: a rural-area coverage analysis. *Vehicular Technology Conference (VTC-Fall), 2017 IEEE 86th*, IEEE, pp. 1–6.

57 Bacco, M., Cassara, P., Gotta, A., and Pellegrini, V. (2019). Real-time multipath multimedia traffic in cellular networks for command and control applications. *2019 IEEE 90th Vehicular Technology Conference (VTC2019-Fall)*, IEEE, pp. 1–5.

12

Airborne Systems and Underwater Monitoring

Elizabeth Basha, Jason To-Tran, Davis Young, Sean Thalken, and Christopher Uramoto

Electrical and Computer Engineering Department, University of the Pacific, Stockton, CA, USA

12.1 Introduction

Wetlands monitoring requires a good understanding of topography and bathymetry. These maps allow for an accurate understanding of the environment. Automation of the map creation process increases the quality of the map and data while reducing personnel time and expenses. Providing that automation using unmanned aerial vehicles (UAVs) reduces the environmental impact as also aerial robots minimize contact time with the environment.

To provide this mapping, we look at a set of challenges and solve them using UAVs. First, the UAVs need to determine where the land and water regions are in a given area. This then allows for specialized robots to map the topography and bathymetry without concerns of topographic aerial robots damaging systems in water and bathymetric aerial robots crashing systems into land.

In order to develop an automated approach to water/land differentiation, the system needs a method to automate the labeling of images. This method combines an algorithm for clustering the image, a hardware system for measuring the points using a UAV, and another algorithm for labeling each image. These images can then provide the training data for a classification system.

We then use the automatically labeled images and develop a system that finds the best machine learning approach to visually differentiating the images taken by the UAV. This is automatically programmed onto the UAV, which can perform online differentiation and determine where the water regions are in GPS coordinates.

Autonomous Airborne Wireless Networks, First Edition.
Edited by Muhammad Ali Imran, Oluwakayode Onireti, Shuja Ansari, and Qammer H. Abbasi.
© 2021 John Wiley & Sons Ltd. Published 2021 by John Wiley & Sons Ltd.

The system then communicates the coordinate information with the UAVs equipped for bathymetric mapping. To start with, we developed an offline approach to determining which points to measure. The offline approach has the point selection performed in advance, then the UAV measures the points, and finally the map is created through interpolation. We simulated this work to determine the best algorithm combination and algorithm parameters for a given number of points to measure.

To improve this method, we explored online approaches to the point selection problem. We implemented and analyzed a set of different algorithms that select the next point to measure based on the current data. Our analysis provides an algorithm that minimizes the root mean square error (RMSE) and maximum difference in the generated map.

These ideas relate to prior work in remote surface classification as well as aerial water measurements. In the area of automatic labeling of images using UAVs, no research has been performed to the best of our knowledge. There is prior work on the remote sensing of wetlands approaches [1–3] that post-process the results and have slower updates due to the increased complexity of satellite and plane data collection. There is also prior work on aerial robot surface classification [4–7]. Similar to the remote sensing work, this work is post-processed with the UAV collecting the data and bringing it back for analysis. Online aerial image processing done by Xu and Dudek focused on coastline detection [8].

To the best of our knowledge, the existing research on online terrain classification by the UAV while flying has focused on using machine learning in support of ground robots. Lunsaeter et al. [9] utilize a Jetson system to provide the online classification; this requires significant power, which limits UAV flight time. Matos-Carvalho et al. [10] use a field-programmable gate array (FPGA). Our work focuses on approaches that can run on a Raspberry Pi, which uses less power (2 W typical [11] compared to 7.5 W typical [12]). Their results are quite quick; however, they require a computer connected to the FPGA for parts of the computation and it is unclear how the system would work online on a UAV.

Finally, there is some work in water sensing using UAVs such as [13]; however, there is none that works to create a bathymetric map using point measurements.

This chapter begins with a discussion of automated image labeling in Section 12.2 followed by describing the approach to differentiating land and water in Section 12.3. Once we know where the water is, we discuss our offline approach to bathymetric mapping using aerial robots in Section 12.4, which also discusses our system. Section 12.5 then outlines our approaches to online bathymetric mapping. Section 12.6 summarizes our current work and outlines our future work.

12.2 Automated Image Labeling

Our first system generates a set of auto-labeled classification images that correctly show where the water is within the image. This system consists of a point selection algorithm, a UAV-based measurement system for testing points, and a region labeling algorithm. It is designed to operate autonomously and online. As the UAV has a limited flight time, it is also designed to be re-entrant in order to allow multiple flights per image (or set of images) as needed to complete labeling.

We first describe the system and then the field experiments testing the system. Note that this builds on our previous publication [14].

12.2.1 Point Selection

Algorithm 12.1 outlines the selection of points. It first reduces the image to a set of clusters based on Hue, Saturation, and Value (HSV) values. A cluster may exist in multiple spatial locations in the image that share similar HSV values; to provide spatial context, we create contours separating the cluster (note that we use cluster to indicate pixels grouped by HSV values and contour to indicate pixels grouped by both HSV value and spatial location). We then examine the contours to find those with areas large enough for the UAV to successfully test, identify the center point of the contours, and select the top 10 points to measure. By selecting only 10 points, we reduce the time needed to measure each image.

Once the points are selected, we order the points from left to right and determine the GPS coordinates of each point.

12.2.2 Measurement System

With the set of points determined, the UAV then flies to each point and measures the conductivity using our custom sensor payload.

The sensor consists of a buoyant cube that sits upright in water, a pair of metal contacts placed on opposite sides of the cube below the natural water line when placed in the water, and a mechanical switch to which the line used to suspend the sensor from the UAV is attached. The metal contacts are placed such that when the sensor is in water, the contacts are submerged, and when the sensor is on land, the contacts are held above the surface. Additionally, the contacts are held above the bottom of the sensor, which helps prevent misclassifications of land as water if the ground is wet or muddy.

The UAV carries the system suspended from a winch and uses the winch to lower the device onto the surface (see Figure 12.1). The device also has a

---

Algorithm 12.1: Point Selection

1: Convert image to HSV
2: Cluster image using K-Means
3: Open morphology
4: Close morphology

5: **for** Each Cluster **do**
6:     Create a bit mask of just that cluster
7:     Find contours of cluster
8: **end for**
9: **for** Each Contour **do**
10:     Determine the contour center point
11:     **if** center > 15 pixels from edge **then**
12:         Use the selected center point
13:     **else**
14:         Test all points to find point furthest from any hole or edge
15:     **end if**
16:     Store selected point with distance and contour area
17: **end for**

18: Sort contours based on distance from selected point to edge
19: Select top 10 points

---

mechanical spring-loaded switch attached between the buoy and the winch. While in the air, the switch is closed due to the tension in the line. Once the buoy device reaches a surface, the tension releases and the switch opens, indicating to the UAV to stop lowering. The buoy then determines that the surface is water if an electrical connection exists or land if no connection exists between the contacts.

12.2.3 Region Labeling

After measuring all points, the system needs to determine the resulting bit mask. The measurements labeled only a subset of the contours so the first need is to label the remaining contours that were not directly measured.

Algorithm 12.2 outlines these steps. It examines the neighboring contours and assigns the label held by the majority of neighbors if more than 25% of the neighbors have a label. If there is no label held by more than 25% of

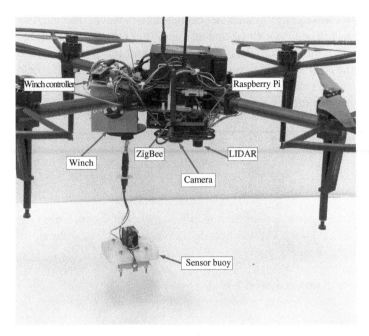

Figure 12.1 Complete system on UAV.

Algorithm 12.2: Region Labeling

1: Given contours and labeled selected points

2: **while** Contours Unlabeled **do**
3:    **for** Each Unlabeled Contour **do**
4:       **if** iterations < 10 **then**
5:          **if** percent neighbor pixels labeled > 25% **then**
6:             Label contour with prevalent neighboring label
7:          **end if**
8:       **else**
9:          Label contour with most prevalent neighboring label
10:       **end if**
11:    **end for**
12:    Increment iterations
13: **end while**

---

its neighbors, the algorithm skips that contour and moves to the next. The algorithm will iterate through all contours until the entire image is labeled or 10 iterations have occurred. If the maximum number of iterations has occurred, the algorithm will select the most prevalent label even if it is held by less than 25% of neighbors. The choice of 25% is to ensure that all contours are labeled; further examination of this parameter is left for future work.

This results in a bit mask for that image, identifying the water and land regions.

12.2.4 Testing

To confirm the system, we first verified the accuracy of the measurement system, simulated the point selection algorithm, and then tested the entire system through field experiments.

12.2.4.1 Measurement System Testing

To test the measurement system, we performed field experiments at the Calaveras River at the University of the Pacific campus in Stockton, California. The river is narrow with significant vegetation, making it similar to wetland environments. It has dirt trails along the banks and roads crossing the river, which we can use to ensure that the system handles man-made features that may be along the sides of wetlands (in California most are near active agriculture regions).

Our experiments used a DJI Matrice 100 system with a custom payload that contains a Raspberry Pi 3, a Garmin LIDAR-Lite v3, a Digi ZigBee radio, and a Logitech C310 webcam. The Raspberry Pi controls all systems, communicating with both the Matrice and any other payloads (such as the winch) to coordinate the experiments.

To start with, we provided the UAV with two points to test: one over water and one over land on the bank of the river. The UAV flew to the first location, lowered the winch, and then lowered itself until the buoy contacted the surface by checking the state of the tension switch. Once the buoy was on a surface, the system took 25 conductivity samples and classified the point. The UAV then flew upward to a set height and repeated the test at that point 10 times. The UAV then raised the winch and flew to the next point to test it.

Overall, the system measured 58 points, 30 water and 28 land. All points were labeled accurately with a 100% success rate, showing that our hardware system can correctly label points.

Table 12.1 Automated point selection simulation results

	Water classification accuracy	False positive percentage	Overall classification accuracy
Mean	96.6	2.9	97.3
Min	47.5	0.0	90.5
Max	100.0	13.4	100.0

12.2.4.2 Point Selection Simulations

With our hardware verified, we simulated the point selection algorithm to provide a larger number of tests.

The simulation used 160 images (gathered from flights over the Calaveras River) to verify the operation of the point selection algorithm. We simulate the measurement phase through manually labeled bit masks that identify water and land regions; we also compare the final result of the algorithm to the labeled bit mask to determine the effectiveness of our system.

Table 12.1 outlines the results of our experiments. On average (looking at the mean values), the system correctly labels 97.3% of all regions, incorrectly labels 3.4% of the water as land (considering 96.6% are correctly labeled as water), and incorrectly labels 2.9% of the land as water. After further analysis of the images, 21 contained no water and were perfectly labeled, providing the best accuracy results.

Figure 12.2 shows the worst accuracy result. As the original image shows (Figure 12.2 a, there is a large contrast within the image between the land on the left side of the image and the right side of the image. Figure 12.2 b

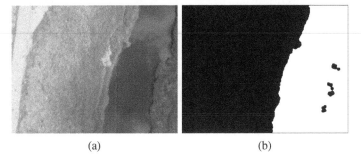

(a)   (b)

Figure 12.2 Worst overall classification accuracy simulation result. (a) Original image. (b) Labeled bit mask.

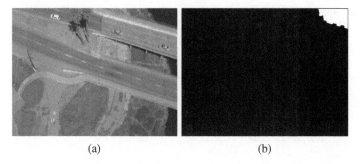

(a)   (b)

Figure 12.3 Worst false negative simulation result. (a) Original image. (b) Labeled bit mask.

shows the system incorrectly labelled most of the land on the right side of the image as water. This result, while not ideal, would train the system to be cautious while measuring as it would ensure that the system sees more water than land, which is safer for our mapping application.

Figure 12.3 shows the worst result in terms of false negatives, where the system incorrectly labels water as land. The image consists of limited water and the water exists right around bridges; this scenario would require a closer, more detailed image to ensure that the UAV does not run into the bridge.

While there are some opportunities for future work in fixing the issues with the worst images, these results are good enough to proceed to field testing the system.

12.2.4.3 Field Experiments

We next implemented the algorithm on the UAV and tested automated point labeling on the Calaveras River. The system flew to a defined GPS location, took a picture, returned to land, and computed the points to measure. Once it had the points, it flew to each point in order, deploying the buoy and measuring at each, until all points were measured. After landing, it completed the process by computing the bit mask.

Figure 12.4 shows the process and result of the experiment. First the UAV took the original image (Figure 12.4a); then it clustered, smoothed, and contoured the image (Figure 12.4b); it selected 10 points to measure (Figure 12.4c); and it constructed the final labeled bit mask of the image (Figure 12.4d). This process required two flights for one image. The winch appeared to be the limiting factor in these experiments; we updated it in our work in Section 12.4.

This proved the overall operation of the system. With labeled images, we next focus on using the images to create a classifier.

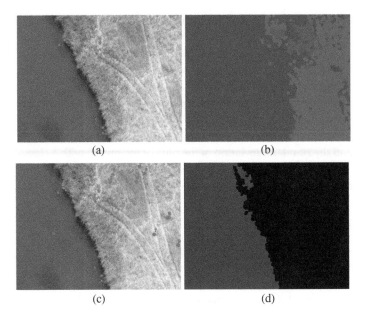

Figure 12.4 Field experiment results for automated point labeling. (a) Original.
(b) Contours. (c) Selected points. (d) Final bit mask.

12.3 Water/Land Visual Differentiation

Once we are able to accurately and autonomously label images using UAVs,
we then use these images to train a machine learning algorithm to differ-
entiate land and water in images. We describe our classifier training and
selection, the online algorithm that performs the complete differentiation,
and our field experiments implementing the system.

12.3.1 Classifier Training

To identify the best classifier algorithm for this problem, we used our
metaclassifier system introduced in [15]. The system generates and trains a
large number of machine learning classifiers with a wide range of features
to select the best for the given data. The set of classifiers focuses on those
requiring less space and computational time.

We modified it to use with images and then ran it on 500 labeled images
(gathered using the system described in Section 12.2.4) using 12 different
algorithms and 12 different features. As mentioned, the algorithm options
focus on less computational time so we chose to not include a combinational

neural network (CNN), despite it performing well on images, in order to see the results we could obtain with simpler algorithms such as decision trees/forests, discriminant analysis, regression, naive Bayes, and similar (see [15] for more details). The feature sets require a minimum of one feature and a maximum of 12, resulting in a large set of models.

We examined the full set of results to determine the best in regard to the highest classification accuracy and the smallest number of features. The best for this problem is the J48 model using two features (the mean of the red in the pixels and the standard deviation of the blue of the pixels) with a classification accuracy of 97.5% and a false positive rate of 2.09. We then used this model to implement on the online classifier.

12.3.2 Online Algorithm

The trained classifier forms a key part of the online algorithm that performs the differentiation. Algorithm 12.3 outlines the steps. After classification, the key elements translate the classified bitmap image into geographic regions and transmit those regions to the other UAVs.

Algorithm 12.3: Online Overview

1: Take picture
2: Measure telemetry (UAV pose and GPS)
3: Classify image using classifier
4: Map Water Regions
5: Transmit Map

12.3.3 Mapping

The mapping algorithm (detailed in Algorithm 12.4) uses a binary classified image that results from our classifier as well as telemetry data pulled from UAV at the time the picture was taken. The binary classified image is a two-dimensional array where pixels classified as water are represented by a one and pixels classified as land are represented by a zero. The binary image is eroded and dilated to remove small spurious land or water classifications within much larger classification areas. This eliminates small classified areas within larger classified areas that are too small to map, which greatly simplifies the map.

Using telemetry data collected from the UAV, we calculate the position and angle at which the picture was taken; this allows us to accurately map each pixel in the image in real world coordinates and offsets. !Specifically,

---

Algorithm 12.4: Mapping Steps

1: Erosion and Dilation
2: Find Contours of Water Regions
3: Map pixels to physical locations
4: Approximate contours within 1 meter of actual contour
5: Filter/eliminate contours based on physical size

---

we use the UAV's pose, represented by a rotational quaternion from a north-east-down (NED) coordinate system, the UAV's height/altitude, measured from take-off, and GPS coordinates. The UAV provides two height /altitude measurements, from take-off and from the WGS-84 ellipsoid, we choose to use the height from take-off as we are interested in the distance from the ground when attempting to map each pixel.

Since wetlands are generally flat, low-lying areas, we can generally assume that there is negligible change in ground altitude across the area surveyed by each flight or that it is within the margin of error introduced by the barometric pressure altitude sensor on the UAV. While altitude measured from the WGS-84 reference ellipsoid may be more consistent across multiple flights, without the use of accurate maps or a ground height database, we would not be able to determine our actual height above the ground. Using all of this telemetry data, we rotate, scale, and transform the image coordinates into real-world coordinates and distances.

Next, we compute the approximate points used to represent the contours of the areas that represent water within a specified safe distance of 1 m of the actual contour. Approximating the contours allows us to reduce the number of points required to represent the area significantly. Additionally, since the contours are approximated within a real-world distance of the actual contour, instead of a pixel distance, the precision remains constant across the entire map across many different pictures taken at different angles and heights.

Finally, we eliminate contours below a specified threshold. This leaves us with physical locations for water within the region provided that are clearly water and large enough for the UAV to work within. We then consider how to communicate this information.

12.3.4 Transmit

Once the contours of the water regions are known for a specified map, we transmit them using the packet structure described in Algorithm 12.5.

Algorithm 12.5: Packet Structure

Number of Contours (N)
for Each Contour **do**
    Number of Points (K)
    GPS Coordinates of First Point
    Offset of each additional point from the first point in whole centimeters (2 * (K-1x))
end for

This results in a packet size of *transmission packet size* $= 9N + \sum 4(K-1)$.

The first byte of the packet specifies the total number contours described by the packet. Next, each contour is described; there is no particular order needed for this description. Each contour is specified by the number points within that contour, the GPS coordinates of the first point in 4-byte float form for both latitude and longitude, and the offset of each remaining point from that first point. By using single precision floats instead of double precision, we incur an error of around 10^{-6} degrees or about 7 cm but reduce our packet size by $8N$ bytes.

This structure balances the amount of communication with the amount of computation the other system has to perform to recreate the region.

12.3.5 Field Experiments

We performed three sets of experiments. First, we calibrated the mapping portion of the algorithm. Second, we verified the operation through simulation to increase the number of test cases. Finally, we performed field experiments validating the entire approach.

12.3.5.1 Calibration

Our initial field experiment calibrated the camera scaling factor used in the calculation of real world position of the map results (based on [16]). We took pictures of the nearby footbridge over the Calaveras and measured the bridge to determine the width of 5.79 m (19 ft). We collected three sets of points per image. These points were converted to real world coordinates using our algorithm and provided a distance; we then calibrated the scaling factor to return distances as close to 5.79 m as possible for our typical flight altitude of 60–80 m. After calibrating the scaling factor, our points now have a mean value of 5.76 m with a standard deviation of 0.24.

12.3.5.2 Simulation

We implemented the algorithm using Python on a Raspberry Pi 3. We ran the algorithm on 195 images to have it calibrate and map the regions.

To verify the differentiation portion, we compared the created map with the original bit mask for the image. On average, the results are 90.9% accurate compared to the bit masks. The best images had an accuracy of 100% and the worst image had an accuracy of 57.1%. Only 13 images were below 80% accuracy, leaving 182 images above that level. Some of these poorer results relate to the classifier accuracy while some relate to the images themselves as some images had more man-made structures (such as the bridge) and less water overall. Overall, though, this experiment confirms the correct operation of the algorithm on the Pi.

We then analyzed the map communication to see how much data was communicated per image. On average the algorithm communicated 22.1 contours or regions of interest. In the worst case, the algorithm identified and communicated 78 contours. This is not unreasonable given the current communication systems although we are exploring ways to compress the information.

12.3.5.3 Overall

Finally, we tested the complete system. We used our DJI Matrice 100 system described in Section 12.2.4 and flew over the Calaveras River on campus. During the flight, the UAV took pictures, classified each picture, and communicated the regions of interest. This successfully demonstrated the operation of our system.

12.4 Offline Bathymetric Mapping

Now that our system can identify the water regions, we focused on using aerial robots to create bathymetric maps. This involves determining the appropriate algorithm for the map creation as well as the hardware to measure depth.

For the algorithm, we analyzed the existing work in bathymetric map creation, dividing the problem into offline point selection and interpolation. We evaluated this first through Matlab simulations and then implemented it in Python on a Raspberry Pi. Through field experiments, we confirmed the basic functionality of the algorithms on the UAV. The initial work in Matlab was previously discussed in [17].

12.4.1 Algorithm Overview

Creating a bathymetric map of an unknown region requires first determining the points to measure, measuring those points, and then interpolating between those points to create the map. As we are starting with an offline approach, the point selection occurs before the UAV takes off and measures. After measurement, the UAV lands and computes the map.

We examined three common methods for point selection when the region is unknown: (i) uniform grid, (ii) random, and (iii) random path. These provide the most reasonable approaches with no prior knowledge of the terrain. All the approaches need the number of points and the region as inputs; random path also has a parameter limiting the maximum distance from the current point to the next point.

To interpolate between the measured points, we explored two different algorithms: (i) inverse distance weighted (IDW) and (ii) spline. IDW examines neighboring values and weights them to create a map based on the assumption that there should not be large discontinuities between neighbors; the algorithm has two parameters defining its behavior: (1) the number of neighbors and (2) the step size in the weighting function. Spline uses radial basis functions (RBFs) to smooth and create the map. The RBFs provide the key parameter for varying the implementation. Our approach used either a multi-quadratic function (MQF) and a multilinear function (MLF).

12.4.2 Algorithm Simulation

To analyze the algorithms and determine the best parameters for our problem, we implemented all of them in Matlab and simulated them using randomized terrains.

We varied the terrains, number of points measured, IDW neighbors, IDW step size, and Spline RBFs. We measured the RMSE and maximum difference between the created map and the original map. Basha et al. provides the details of these tests [17].

After exploring the impact of all these parameters, we determined the best offline approach based on the number of points to measure and the preferred metric. For the point selection, uniform grid provided the best results under all cases.

Table 12.2 shows the results of the interpolation algorithm analysis. Spline provides the best overall, but the parameters differ. If the number of points is less than 25, the metric chosen matters. To have the lowest RMSE, spline with a quadratic basis function and constant of 2.5 provides the best interpolation; however, for reducing the maximum difference, a linear basis function and constant of 1 provides the best result. For points greater than and

Table 12.2 Best case interpolation decision table

Sample points	RMS error	Maximum difference
< 25	MQF, 2.5	MLF, 1
≥ 25	MQF, 5	MQF, 5

equal to 25, a quadratic basis function with a constant of 5 provides the best results independent of the metric.

12.4.3 Algorithm Implementation

Once simulated, implementation of the algorithm commenced. Implementing the algorithm requires additional considerations in terms of the overall flow of the system and how the algorithm connects with the UAV.

The system is designed to have a Raspberry Pi 3 (the Pi) control all behavior; the Pi is a good embedded system for controlling UAVs and other hardware. The first step was translating the Matlab into Python to run on the Pi; we used Python 3 and NumPy to ensure equivalent functionality. In order to verify that our Python code produced output that was identical to that of the Matlab code, the Matlab terrains were saved into readable files, equal tests were performed on both codebases using those terrains, and the outputs were compared to ensure identical results.

Once the algorithms worked correctly, the system expanded to connect to the UAV. For this system, a DJI Matrice 100 was used as this UAV provides software control points for external automation. DJI provides a software developer kit with an application programming interface (API) to control and to communicate with the Matrice.

The Pi runs the bathymetric point selection algorithm to start. The system then runs through each point with the Pi sending movement commands to the UAV; these points can be provided in relative or GPS coordinates to allow for maximum flexibility. Once the UAV communicates that it has reached the point, the Pi controls the measurement system to measure the depth at that point. One nuance of the implementation is the need for multiple flights to measure all points. The UAV has a limited flight time (around 15–20 minutes depending on wind conditions). Therefore, the system allows for recording a selection of points, recharging the UAV, and re-entering the process for the next set of points. Once all points are measured, it creates the map through the interpolation algorithm.

The system algorithms, movement, and battery management were confirmed to work through field experiments without the measurement system. The system ran through a number of point selection waypoints, moving correctly to the start of the region and then to each point in order. The system could correctly stop measurements, return to base, and restart in the correct location as well.

12.4.4 Bathymetric Measurement System

With the software developed and implemented, the next focus is on the measurement hardware system. This attaches to the UAV as a payload and works with the UAV to measure water depth. It consists of two parts: the winch that directly attaches to the UAV and the depth measurement payload that connects to the winch. This section discusses both parts as well as the testing.

The winch system consists of the Raspberry Pi, two motors to raise and lower the cable, a custom-designed board, and a battery (as shown in Figure 12.5). Two motors were selected in order to increase the speed of the winch system, which allows for more measurements to occur in any one flight and larger payloads. The custom board uses a Teensy LC microcontroller to control the motors and communicate via serial to the sensor payload. The winch has a limit of a 150 g payload. With a 126 g payload, the average time to lower the payload is 4 seconds and to raise the payload is 6.1 seconds.

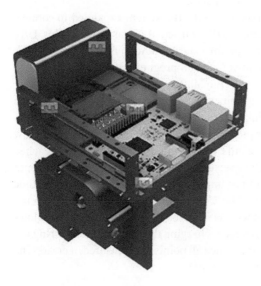

Figure 12.5 Winch and Raspberry Pi.

Figure 12.6 Sensor payload.

The sensor payload performs the depth measurements; note that this is a different sensor payload than that used for automatic labeling of points. Shown in Figure 12.6, it consists of a JSN-SR04T depth sensor, an Arduino Pro Micro microcontroller, and a battery. The JSN-SR04T is a waterproof sensor consisting of an ultrasonic transmitter/receiver pair; the system measures depth by measuring the pulse length response in microseconds from the sensor. When requested by the winch system, it sends the measurement depth as a float value to capture the details of the timing measurement.

This ultrasonic depth sensor was tested in depths ranging from 0.9 to 4 m in both a controlled pool environment and the nearby Calaveras River. From the testing results, it was determined that the sensor had a minimum testing range of 0.9 m when in water. Unfortunately, complete system testing connecting the hardware with the bathymetric map software was interrupted by the COVID-19 pandemic. This will commence as soon as possible.

12.5 Online Bathymetric Mapping

Our offline approach does not take full advantage of the adaptability that our UAV provides. Once we confirmed that we could use UAVs for bathymetric mapping, we furthered our work by exploring online approaches that adapt to the information acquired. To do so, we determined a number of point selection algorithms to use based on the existing problems in robotics and coverage space.

Our selection criteria for evaluation included computational complexity while we ran simulations to evaluate our approaches for accuracy and other issues. For our simulations, we implemented all algorithms in python within our existing software framework and combined the point selection algorithms with our existing interpolation algorithms.

In this section, we discuss the algorithms we evaluated, our evaluation methodology, and the results.

12.5.1 Point Selection Algorithms

We selected six different algorithms to evaluate: (i) Monotone Chain Hull, (ii) Incremental Hull, (iii) Quick Hull, (iv) Gift Wrap, (v) Slope-based, and (vi) Combination. All algorithms start with a maximum number of points to measure and an outline of the area to map; each returns the set of measured coordinates and measured depths. Each algorithm is described in the following subsections.

12.5.1.1 Monotone Chain Hull Algorithm

Monotone Chain Hull begins by bisecting the region in two and initially distributing the points equally over the bisection. It then starts with the first column and randomly selects the upper or lower hull. The algorithm measures at that location and in the direction of the hull (upper or lower) until the slope stabilizes and stops increasing or it has reached the upper or lower bounds of the region. It continues this process with each of the points until the hull has been created. We utilize all points measured to create our map.

12.5.1.2 Incremental Hull Algorithm

The Incremental Hull algorithm begins by creating an initial hull in the center of the map. The center hull has a defined radius from the center of the overall region; the initial set of points are equally spaced on this center hull and measured.

Once the algorithm measures the points from the initial hull, it will begin the traversal. The traversal begins by selecting any point from the initial hull. The algorithm will measure from that point outward at defined step sizes until the slope stabilizes or it reaches an edge of the region. The algorithm then selects a different point.

12.5.1.3 Quick Hull Algorithm

The Quick Hull begins by performing a diagonal traversal across the map. Upon completing that initial traversal, the algorithm will then randomly select the portion of the map to traverse (above or below the diagonal). From there, the algorithm will begin selecting points along the edge of the map and traversing to the point. A finite state machine determines the offset size between the incoming points based on the terrain and the rate of change of the slope; areas of limited change are traversed faster than areas of significant change. Once reaching the selected point, the algorithm will reset to the starting point, select a new point, and then traverse again.

12.5.1.4 Gift Wrap Algorithm

Gift Wrap starts by randomly dispersing points throughout the region using a uniform probability. It then selects the point farthest to the left of the region and measures it. From that point, the algorithm selects a point closest to it on the right and above it (if multiple points exist). This point is measured. Selecting the next measurement point involves calculating the angles between these last two points measured. It continues measuring and finding points of the largest angle until the maximum number of points to measure has been reached.

12.5.1.5 Slope-Based Algorithm

The Slope-based algorithm is a point selection algorithm that follows the direction of the steepest slope. It attempts to explore areas that have more information based on the idea that areas where the slope is changing more quickly are more informative. The algorithm begins with an initial set of eight measurements starting at the upper left of the region and moving a random distance through an ordered list of the cardinal directions (E, S, W, SE, NE, SW, and N). With all the measurements, the pairwise slopes are calculated between the points. The algorithm then determines a random distance to travel and moves through that distance in the direction of the steepest slope (as long as it remains within the region). It will update the slope measurements and continue to measure in whatever direction the steepest slope is until the maximum number of measurements is reached.

While easy to compute, as expected, one issue with this type of algorithm is that it often gets stuck in local minima. Attempting to alleviate this leads to the next algorithm.

12.5.1.6 Combination (Slope-Based and Probability) Algorithm

In order to alleviate the local minima issue, the Combination algorithm adds a probability aspect. Instead of only focusing on the slope, the algorithm randomly chooses a new direction to follow. Initially this random direction parameter is set to 50%, providing an equal chance of choosing to follow the slope-based approach or to explore; depending on a threshold parameter, the probability is incrementally changed to increase randomness or increase the slope following approach. The threshold parameter is initially equally balanced between the two options and shifts as the overall slopes change.

12.5.2 Simulation Setup

All algorithms were implemented and tested in Python 3.

Overall, the simulations varied the number of points to measure, the terrain, and algorithm-specific variables. The tests used 25 as the number of points to measure as the terrains were set to 25 by 25 grid elements. This provides a reasonable comparison to the number of points that the system can measure relative to the region size. For the terrains, several baseline terrains were created: one that provided a smooth channel and one that provided a constant slant. Additionally, two random terrains were created.

All combinations of point selection algorithms and interpolation algorithms were tested. For the point selection algorithms, only incremental hull has a parameter to vary, the step size. This parameter was varied to be 1, 2, or 3. We additionally tested the uniform grid point selection algorithm from the offline approach to see how our online approaches performed.

Each of the interpolation algorithms had parameters to modify. Spline has the RBF and the RBF weight (c). There are five different RBFs to test: MQF, MLF, inverse multi-quadratic function (IMQF), natural cubic splines function (NCSF), and thin plate splines function (TPSF). The function weight, c, was varied to cover all values in the set (0.5, 1, 2, 2.5, 5, 10, 20). The IDW algorithm has the exponent to vary; our tests used 1 or 2 as this value. The neighbors are also varied in a range from 1 to 8.

For every setting combination, the test is run 10 times and averaged. The generated terrain is compared to the initial terrain using the same two metrics as our offline work: RMSE and maximum difference.

12.5.3 Results and Analysis

We analyzed the results for each interpolation algorithm before determining the best combination for our online bathymetric mapping.

12.5.3.1 Spline

We first examine the Spline interpolation algorithm. In regard to RMSE, the results are consistently better with the MQF or MLF basis function, echoing our offline results. For maximum difference, IMQF provides the best results.

The algorithm performs better with lower values of C (below 10) as it gets exponentially worse above 10. There appears to be a linear relationship between the constant values and RMSE/maximum difference, where

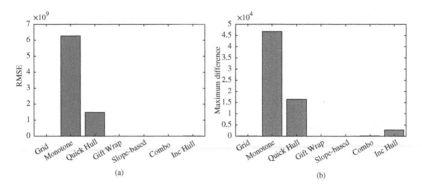

Figure 12.7 Spline results. (a) RMSE. (b) Maximum difference.

they are minimized with the smaller constant values. The exception to this is a constant of 10, which minimizes both the RMSE and the maximum difference.

The results for both RMSE and maximum difference are shown in Figure 12.7. In regard to the point selection algorithm, Grid or Gift Wrap performs the best in regard to RMSE. Gift Wrap or Slope-based performs the best in regard to maximum difference. Monotone Chain performs significantly worse than all other algorithms.

12.5.3.2 IDW

We next examine the IDW interpolation algorithm. In terms of parameters, for both metrics, the results are consistently better with an exponent of 2 and the number of neighbors between 3 and 5.

The results for both RMSE and maximum difference are shown in Figure 12.8. Across the settings, Gift Wrap performs the best for both

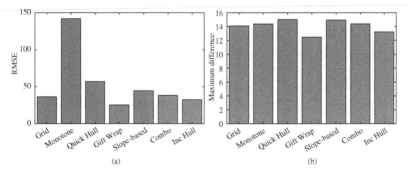

Figure 12.8 IDW results. (a) RMSE. (b) Maximum difference.

metrics as well followed by Incremental Hull. Monotone Chain continues to perform the worst although not at the degree seen with Spline.

12.5.3.3 Overall Summary

Looking at all algorithms and settings, the best choice is the Gift Wrap point selection algorithm combined with the IDW interpolation algorithm.

We implemented this within our existing Python code base to reduce the time for field experiments. Running a field experiment with these algorithms and the hardware described in Section 12.4.4 is the next step once it becomes possible.

12.6   Conclusion and Future Work

Our work provides a set of systems using UAVs to perform bathymetric mapping. It determines where the water region of interest is automatically based on aerial images and then creates a map of that region. Depending on timing and processing, that map could be based on an offline point selection algorithm of uniform grid or an online point selection algorithm of gift wrap paired with the best interpolation algorithm for each and with all parameters determined. We design a custom winch and sensing system to attach to the UAVs to perform the needed depth measurement.

In the future, once field experiments can resume, we plan to fully test the online algorithms and connect all pieces into a cohesive field experiment. We are also examining methods of sharing the computation of the classification approach in order to try more complicated neural network algorithms.

Bibliography

1 Baker, C., Lawrence, R., Montagne, C., and Patten, D. (2006). Mapping wetlands and riparian areas using Landsat ETM+ imagery and decision-tree-based models. *The Society of Wetlands* 26 465, https://link.springer.com/article/10.1672/0277-5212(2006)26[465:MWARAU]2.0.CO;2.

2 Martyn, R.D., Nobel, R.L., Bettoli, P.W., and Maggio, R.C. (1986). Mapping aquatic weeds with aerial color infrared photography and evaluating their control by grass carp. *Journal of Aquatic Plant Management* 24 46–56, https://www.cabi.org/ISC/abstract/19860788995.

3 Ramachandra, T.V. and Kumar, U. (2008). Wetlands of greater Bangalore, India: automatic delineation through pattern classifiers. *Electronic Green Journal* (26), https://escholarship.org/uc/item/3dp0q8f2#author.

4 Casado, M.R., Gonzalez, R.B., Kriechbaumer, T., and Veal, A. (2015). Automated identification of river hydromorphological features using UAV high resolution aerial imagery. *Sensors* 15 (11): 27969–27989, https://www.mdpi.com/1424-8220/15/11/27969.

5 DeBell, L., Anderson, K., Brazier, R.E. et al. (2015). Water resource management at catchment scales using lightweight UAVs: current capabilities and future perspectives. *Journal of Unmanned Vehicle Systems* 4 (1): 7–30, https://cdnsciencepub.com/doi/10.1139/juvs-2015-0026.

6 Hung, C., Xu, Z., and Sukkarieh, S. (2014). Feature learning based approach for weed classification using high resolution aerial images from a digital camera mounted on a UAV. *Remote Sensing* 6 (12): 12037–12054.

7 Papakonstantinou, A., Topouzelis, K., and Pavlogeorgatos, G. (2016). Coastline zones identification and 3D coastal mapping using UAV spatial data. *Unmanned Aerial Vehicles in Geomatics* 5: 75, https://www.mdpi.com/2220-9964/5/6/75#cite.

8 Xu, A. and Dudek, G. (2010). A vision-based boundary following framework for aerial vehicles. *Proceedings of the IEEE/RSJ International Conference on Intelligent Robots and Systems (IROS)*.

9 Lunsaeter, S.F., Iwashita, Y., Stoica, A., and Torresen, J. (2020). Terrain classification from an aerial perspective. *2020 IEEE International Conference on Systems, Man, and Cybernetics (SMC)*, pp. 173–177.

10 Matos-Carvalho, J.P., Moutinho, F., Salvado, A.B. et al. (2019). Static and dynamic algorithms for terrain classification in UAV aerial imagery. *Remote Sensing* 11 (21), https://www.mdpi.com/2072-4292/11/21/2501#cite.

11 Raspberry Pi Organization. (2020). Raspberry pi power requirements. https://www.raspberrypi.org/documentation/hardware/raspberrypi/power/README.md (accessed 13 March 2021).

12 NVIDIA. (2017). NVIDIA Jetson TX2 Delivers Twice the Intelligence to the Edge. https://developer.nvidia.com/blog/jetson-tx2-delivers-twice-intelligence-edge/ (accessed 13 March 2021).

13 Ore, J.-P., Elbaum, S., Burgin, A., and Detweiler, C. (2015). Autonomous aerial water sampling. *Journal of Field Robotics* 32 (8): 1095–1113.

14 Klein, C., Speckman, T., Medeiros, T. et al. (2018). UAV-based automated labeling of training data for online water and land differentiation. In: *Proceedings of the 2018 International Symposium on Experimental Robotics*, ISER 2018, Buenos Aires, Argentina (5–8 November 2018), *Springer Proceedings in Advanced Robotics*, vol. 11 (ed. J. Xiao, T. Kroger, and O. Khatib), 106–116. Springer.

15 Basha, E., Watts-Willis, T., and Detweiler, C. (2017). Autonomous meta-classifier for surface hardness classification from UAV landings. *2017 IEEE/RSJ International Conference on Intelligent Robots and Systems*, IROS 2017, Vancouver, BC, Canada (24–28 September 2017), IEEE, pp. 3503–3509.

16 Barber, D.B., Redding, J.D., McLain, T.W. et al. (2006). Vision-based target geo-location using a fixed-wing miniature air vehicle. *Journal of Intelligent and Robotic Systems* 47 (4): 361–382.

17 Basha, E., Morales, C., Thalken, S. et al. (2020). Automated bathymetric mapping using unmanned aerial vehicles for wetlands monitoring. *2020 International Congress on Environmental Modelling and Software*, iEMSs 2020, Brussels, Belgium (14–18 September 2020).

13

Demystifying Futuristic Satellite Networks: Requirements, Security Threats, and Issues

Muhammad Usman[1], Muhammad R. Asghar[2], Imran S. Ansari[3], and Marwa Qaraqe[1]

[1]College of Science and Engineering, Education City, Division of Information and Computing Technology, Hamad Bin Khalifa University (HBKU), Doha, Qatar
[2]School of Computer Science, The University of Auckland, Auckland, New Zealand
[3]James Watt School of Engineering, University of Glasgow, Glasgow, UK

13.1 Introduction

Fifth generation (5G) of cellular networks promises to make our lives easier, safer, and healthier by providing connectivity to different vertical industries such as self-driving cars, telemedicine, Internet-of-things (IoT), and smart cities. To make this dream true, 5G promises to provide 10 Gbps speed with a milliseconds latency. However, new use cases, such as the desire to connect the unconnected 3.9 billion people, are expected in 6G networks [1]. To connect remote locations with a cellular network, satellite networks seem to be the most viable solution. Basically, satellite networks are complementary to terrestrial networks in order to provide coverage to remote rural areas [2]. However, in satellite networks, there is a lack of support for many useful use cases (high throughput and low latency), which are inherently supported by terrestrial networks.

Because of the aforementioned reasons, satellite networks have gained much attention in recent decades. Indeed, an Integrated Satellite and Terrestrial Network (ISTN) has been proposed in the literature for future cellular networks (e.g. 6G and beyond) to cover a broad range of use cases alongside global coverage [3]. This integration will provide higher data rate coverage to areas where conventional cellular networks are either unable to reach or damaged (partially or completely) due to disasters.

Apart from providing global Internet coverage, the futuristic satellite networks will provide access to interplanetary communication, which is

Autonomous Airborne Wireless Networks, First Edition.
Edited by Muhammad Ali Imran, Oluwakayode Onireti, Shuja Ansari, and Qammer H. Abbasi.
© 2021 John Wiley & Sons Ltd. Published 2021 by John Wiley & Sons Ltd.

necessary to not only explore the solar system but also to connect deep space satellites to the Earth. In this regard, the future satellite networks will be highly dynamic in nature, ranging from ground stations (GS) on the Earth to Interplanetary network including satellites on deep space exploration missions. We call this complete ecosystem of satellite network as inter-satellite and deep space network (ISDSN) [4].

Such a dynamic environment poses different challenges not only in terms of networking, inter-satellite links (ISL), routing, and physical layer communication paradigms but also in terms of security of those networks. All the security challenges faced by terrestrial networks apply to ISDSN as well. Note that the eavesdropping of a communication link becomes extremely difficult in a deep space environment. However, the possibility cannot be neglected due to the introduction of nano- and pico-satellites, such as Cube-Sats. Owing to their low deployment cost and development of miniaturized thrusters, CubeSats and ultra-small satellites are considered to potentially revolutionize the deep space industry [5]. However, due to their cheap development cost, a malicious entity may launch a satellite in space for illegitimate purposes, such as eavesdropping or data tampering, etc. It is necessary to incorporate security as an essential aspect during the design and configuration of an ISDSN at all layers.

To address security challenges faced by an ISDSN, physical layer security can be a potential solution [6]. The physical layer paradigm should be wisely chosen from among radio frequency (RF), free-space optical (FSO) communication [7–10], and/or millimeter Wave (mmWave) [11]. Owing to the highly directional beam characteristics of FSO communications, it is very difficult to eavesdrop on the communication link [12].

In this work, we classify the ISDSN into different tiers highlighting the communication and networking paradigms. In addition, we discuss the security requirements, challenges, and threats at each tier. The rest of the chapter is divided into the following sections. Section 13.2 presents a three-tier classification of an ISDSN including interplanetary networks. Security requirements, challenges at different tiers of an ISDSN, and their potential solutions are described in Section 13.3. Finally, Section 13.4 concludes the chapter.

13.2 Inter-Satellite and Deep Space Network

Based on the communication, futuristic satellite networks could be divided into three tiers. Each tier is different in terms of communication environment, communication protocols, types of communication nodes, and its maturity level (i.e. is the network currently operational or not?).

13.2.1 Tier-1 of Satellite Networks

Tier-1 includes the communication of a satellite with the GS. This is the most conventional satellite network, and has been in operation for decades. One of the traditional uses of tier-1 of satellite networks is as a backup to fiber-optics backhaul links for long-range communications [13]. In addition, these networks are also utilized to provide coverage to remote areas where cellular or wired networks are not available. The use cases at this tier of satellite networks can be mainly divided into three types: (i) urban scenarios; (ii) rural scenarios; and (iii) remote scenarios [14] (see Figure 13.1).

Urban use cases include, but are not limited to, providing backhaul to a number of scenarios, such as media broadcast, 5G base stations, commercial IP backhaul, and providing access to various IoT devices. Similarly, rural use cases include providing connectivity to areas where conventional cellular networks are not available. Satellite networks can be used to extend cellular coverage in rural areas. Remote use cases include providing connectivity to deep-sea ships, airplanes, disaster relief vehicles operating in a remote area, and to personnel working in an extremely remote location, such as oil and gas exploration in a deep desert.

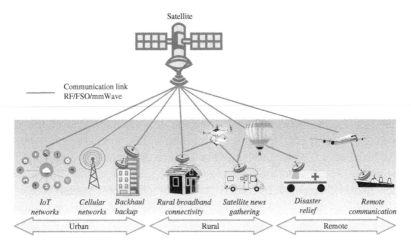

Figure 13.1 Tier-1 of satellite networks: It includes the connection of a satellite with the ground station.

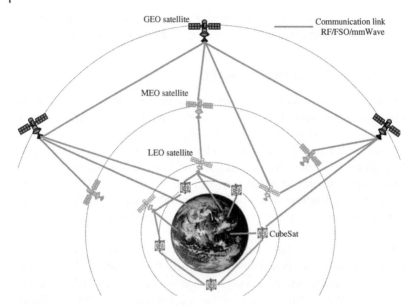

Figure 13.2 Tier-2 of satellite networks: It includes the inter-satellite links between GEO, MEO, and LEO satellites.

For most of the aforementioned scenarios, the communication protocols at this tier are well-defined along with the security mechanisms. Mostly at this tier, RF is utilized for communication. However, recent technologies, such as FSO [15] and mmWave [16], are also being investigated for use in tier-1 networks. In addition, the recent trend of using satellite networks to provide connectivity to IoTs [17] has posed serious security challenges across the industry. These security challenges are discussed in Section 13.3.

13.2.2 Tier-2 of Satellite Networks

Tier-2 of satellite networks include the ISL between various satellites in low earth orbit (LEO), medium earth orbit (MEO), and geosynchronous equatorial orbit (GEO). A vision of tier-2 of satellite networks is presented in Figure 13.2. The potential use cases at the tier-2 of satellite networks include providing global Internet coverage using multiple constellations of satellites. To date, multiple constellations of satellite exist in different orbits (LEO, MEO, GEO) wherein all the satellites of a constellation are connected through communication links. The details of satellite constellations launched in space to date can be found online.[1]

---

1 NewSpace Index: https://www.newspace.im.

It is important to note that most of the modern constellations sent in space have the capability of establishing ISL either using FSO or RF. For instance, *Telesat LEO* [18], an initiative of Airbus, is a LEO constellation of 117–512 satellites with the ability to establish inter-satellite optical links. Similarly, the Starlink LEO satellite constellation from SpaceX[2] has the provision of optical ISL.

Most of the constellations utilize their proprietary ISL protocols. For instance, Iridium satellites in space utilize RF as a communication means and utilize Motorola proprietary ATM-like switching as an ISL protocol [19]. Moreover, Telesat has the provision of beam-steering and shaping for at least 16 beams with the ability of IP-routing [20].

13.2.3 Tier-3 of Satellite Networks

Tier-3 of satellite networks includes the communication beyond GEO orbits. It includes the communication of the Earth (or tier-2 satellites) with the moon, the planets, and deep space satellite missions. A vision of satellite networks at this tier is given in Figure 13.3. The concept of interplanetary communication and Internet services has emerged in recent decades [21]. Although many options for communication spectrum are available at this tier (e.g. RF, FSO, or mmWave), FSO may become the more preferred option due to substantial availability of the spectrum [22, 23]. This is evident from the National Aeronautics and Space Administration (NASA)'s first space laser communication system demonstration [24], which utilizes lasers as a transmission technology to establish a link between moon and the Earth able to communicate a maximum rate of 622 Mbps. This demonstration has laid the groundwork for future optical relay networks in space for interplanetary communication.

In terms of routing and networking protocol, this kind of networks with such a long distance between communicating nodes may be handled by the concepts of a delay-tolerant network (DTN) [25]. DTN has emerged as an approach to solve the problems that cannot be handled by typical routing protocols. One approach, for instance, can be to utilize a "store and forward" technique, wherein data is essentially stored at each hop and incrementally moved forward [26].

Apart from the interplanetary Internet, the communication of satellites on deep space exploration mission with the Earth is another important element of tier-3 of satellite networks. These satellites may act as relays to connect the Earth with other planets.

---

2 https://www.spacex.com.

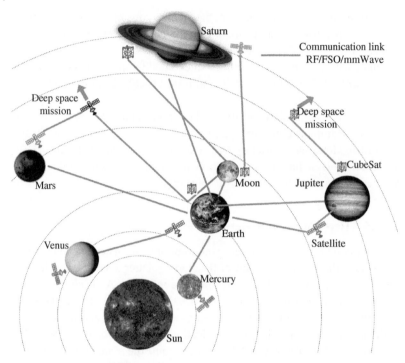

Figure 13.3 Tier-3 of satellite networks: It includes the communication of the Earth with different planets, moon, and deep space satellite missions.

13.3 Security Requirements and Challenges in ISDSN

The overall security requirements of ISDSN remain the same at all tiers. However, depending on the satellite/node's capabilities, the physical communication paradigm (RF/FSO/mmWave) and their location in the ISDSN, potential security threats change at every tier of the network. For instance, the communication paradigm at tier-3 of satellite networks is different from that at tier-1 and tier-2 wherein the average distance between communicating nodes is extremely large, as is their distance from the Earth. It becomes potentially difficult to eavesdrop on any interplanetary communication at this level. However, the chances of being compromised by a node at the Earth, in scenarios wherein the Interplanetary network becomes a part of the Earth's Internet, cannot be neglected.

In what follows, we first enlist the general security requirements of an ISDSN at all tiers and discuss some security threats and challenges at all tiers.

- *Confidentiality*: The data must be exposed to authorized entities only. Each entity must prove its identity by means of an authentication mechanism. Furthermore, confidentiality must be ensured at all network nodes, while the data is at rest (i.e. buffer, storage, etc.) and in transmission.
- *Integrity*: The data must not be modified by any unauthorized entity. Moreover, the source of the data (i.e. data provenance) must be verifiable.
- *Availability*: The authorized entities should be able to access the device and the data at all times.

The rest of this section discusses the aforementioned security requirements at various tiers of an ISDSN.

13.3.1 Security Challenges

The security challenges faced by an ISDSN change with the communication environment, node capacities, the distance between nodes, and how easy it is to eavesdrop on the communication between two nodes [27]. Clearly, the network devices at tier-1 are more prone to cyberattacks followed by tier-2 because of easy access to threat actors. At tier-3, the network devices are quite far from the Earth (in deep space and solar planets) but the chances of getting compromised are not void. In what follows, we will discuss some security challenges at each tier of an ISDSN.

13.3.1.1 Key Management
Key management in an ISDSN network remains an open research challenge, especially at tier-2 and tier-3 of satellite networks. Tier-1 has been in operation for many decades and the key management issues are thoroughly discussed in the literature [28, 29]. For instance, the work presented in [29] elaborates on the dynamics of key management in satellite multi-cast networks. However, key management at tier-2 and tier-3 becomes more complex due to certain issues, such as longer propagation delays, extremely fast mobility of satellite nodes in space, and frequent handovers. In this highly dynamic scenario, it becomes very critical to choose suitable cryptography-based key management technologies. Key management has become an essential part of communication at almost every layer. For geographically large areas, it becomes difficult to have an online key

management center wherein public keys and symmetric keys are efficiently managed across an inter-satellite or deep space network.

In the literature, key management techniques are generally classified into four categories [30]: (i) centralized; (ii) distributed; (iii) topology-based; and (iv) pre-configured. All of those may have some advantages and disadvantages in the framework of tier-2 and tier-3 of ISDSN. For instance, if a scenario of a node joining or leaving is considered, the centralized key management seems to be the most viable option, which makes it easier to update the shared keys. However, in an ISDSN environment, it is very difficult to have a central management authority due to the extremely vast geographic distribution of network nodes. Moreover, due to the highly dynamic environment of ISDSN nodes joining/leaving, frequency may be quite high so a key management protocol must be designed keeping in view the dynamics of such a network.

Another important aspect is higher bit error rates (BER) in satellite environments, which may arise from poor line of sight (LoS) in case of FSO links or higher number of interfering nodes in case of RF. The higher BER, higher communication load, and higher time delays restrict the choice of an efficient key management protocol. Key management in such an environment must primarily focus on reducing the round complexity and improving the rekeying efficiency, especially for the scenarios with new joining/leaving.

13.3.1.2 Secure Routing

The routing protocols in a classical satellite network at tier-1 are broadly studied in the literature [31]. However, the routing between ISL and in the interplanetary network at tier-2 and tier-3, respectively, are least addressed in the literature. Given the distance between nodes at ISL and deep space, it becomes quite challenging to have secure routing protocols to obtain connectivity between all nodes. The situation has become more complex, due to the launch of many CubeSats not only in LEO but also for deep space mission. To include those into the satellite network ecosystem, efficient routing protocols are required for tier-2 and tier-3 of ISDSN.

Owing to the highly dynamic environment at tier-2 and tier-3, the routing protocol must consider the following aspects, which also directly or indirectly affect the security of satellite networks. First of all, the mobility of nodes must be considered. The routing protocol must adapt with the dynamics of the network ensuring an acceptable blocking probability during handovers. Second, to minimize link delays, the routing path must be optimized, reducing the number of hops. Third, the protocol must balance the load among various satellites if different routes are possible.

Last, but not least, the protocol must employ a robust security mechanism against various attacks.

There exist some works in the literature that focus on routing in satellite networks. However, most of them focus on tier-1 only. A survey of such protocols is provided in [32]. It can be observed from [32] that the literature of secure routing protocols in satellite settings is not exhaustive. Further, the routing at tier-2 and tier-3 is rarely studied in the literature, let alone the security issues in those protocols.

Apart from the aforementioned security challenges, some other challenges in an ISDSN network can be summarized as scalability of the satellite security protocols and authentication mechanisms for satellite entities.

13.3.2 Security Threats

In what follows, we briefly discuss the potential security threats at all tiers of an ISDSN.

13.3.2.1 Denial of Service Attack

Denial of Service (DoS) attacks in satellite networks is to block the communication with a network node through flooding or through any other means, thereby compromising the availability of the network to legitimate users. Similar to any terrestrial network, satellite networks are also prone to DoS and widespread distributed denial of service (DDoS) attacks. Although some works in the literature discuss DoS attacks at tier-1 of ISDSN, the possibilities of DoS and DDoS attacks at tier-2 and tier-3 are rarely discussed in the literature. Particularly, the possibilities of DoS attacks at tier-3 of an ISDSN are not discussed in the literature. One interesting work regarding DoS attack in satellite networks is presented in [33, 34]. Such works focus on only tier-1 of ISDSN. However, the techniques used to mitigate DoS and DDoS attacks [35] can be extended to tier-2 and tier-3.

13.3.2.2 Data Tampering

Data tampering in ISDSN refers to modifying, manipulating, and/or editing sensitive information, thereby compromising the integrity of the network. Owing to the broadcast nature of satellite network at tier-1, it is easier for an eavesdropper to overhear the satellite communication and potentially modify it. However, tier-2 and tier-3 are difficult to be compromised although the chances cannot be neglected. It is quite possible that an adversary may launch a satellite, such as a CubeSat, with a malicious intention to eavesdrop on the communication at tier-2 and tier-3 and potentially tamper with the communication.

A possible solution to the aforementioned security issues, such as key management, data tampering, etc. is to utilize physical layer security at satellites [36], especially for the scenarios wherein they are connected via FSO links [37]. FSO itself is a more secure technology than its counterparts, such as RF. Unlike RF, FSO-based communication mainly occurs through very converged beam [38] wherein it becomes very difficult for an eavesdropper to overhear any information.

Another emerging solution to secure the long-distance communication system is quantum key distribution (QKD) wherein a secret is shared between two communicating parties using a cryptographic protocol involving components of quantum mechanics [39]. In this regard, satellite-based QKD systems have been proposed to establish secure intercontinental links [40].

13.4 Conclusion

This chapter classified futuristic satellite networks into three tiers, ranging from a GS on the Earth to interplanetary and deep space networks. We believe that such networks will be an essential part of future cellular networks (say 6G and beyond), wherein the satellites in LEO orbits will play a key role in providing global coverage. At the same time ISL and deep space networks will connect the complete solar system together with the Earth acting as the heart of the Internet. We highlighted the communication and networking paradigms at each tier. Finally, we discussed security requirements, research challenges, and threats of such dynamic networks and their potential physical layer solutions.

Bibliography

1 Yaacoub, E. and Alouini, M. (2020). A key 6G challenge and opportunity-connecting the base of the pyramid: a survey on rural connectivity. *Proceedings of the IEEE* 108 (4): 533–582.

2 Yang, L. and Hasna, M.O. (2015). Performance analysis of amplify-and-forward hybrid satellite-terrestrial networks with cochannel interference. *IEEE Transactions on Communications* 63 (12): 5052–5061.

3 Huang, X., Zhang, J.A., Liu, R.P. et al. (2019). Airplane-aided integrated networking for 6G wireless: will it work? *IEEE Vehicular Technology Magazine* 14 (3): 84–91.

4 Saeed, N., Elzanaty, A., Almorad, H. et al. (2020). Cubesat communications: recent advances and future challenges. *To appear in IEEE Communications Surveys Tutorials* 22 (3): 1839–1862.

5 Levchenko, I., Keidar, M., Cantrell, J. et al. (2018). Explore space using swarms of tiny satellites.

6 Illi, E., Bouanani, F.E., Ayoub, F., and Alouini, M.-S. (2020). A PHY layer security analysis of a hybrid high throughput satellite with an optical feeder link. https://arxiv.org/abs/2003.12358.

7 Ai, Y., Mathur, A., Cheffena, M. et al. (2019). Physical layer security of hybrid satellite-FSO cooperative systems. *IEEE Photonics Journal* 11 (1): 1–14.

8 Yang, L., Gao, X., and Alouini, M. (2014). Performance analysis of free-space optical communication systems with multiuser diversity over atmospheric turbulence channels. *IEEE Photonics Journal* 6 (2): 1–17.

9 Ansari, I.S., Yilmaz, F., and Alouini, M. (2013). On the sum of squared η-μ random variates with application to the performance of wireless communication systems. *2013 IEEE 77th Vehicular Technology Conference (VTC Spring)*, pp. 1–6.

10 Ansari, I.S., Abdallah, M.M., Alouini, M., and Qaraqe, K.A. (2014). A performance study of two hop transmission in mixed underlay RF and FSO fading channels. *2014 IEEE Wireless Communications and Networking Conference (WCNC)*, pp. 388–393.

11 Vuppala, S., Tolossa, Y.J., Kaddoum, G., and Abreu, G. (2018). On the physical layer security analysis of hybrid millimeter wave networks. *IEEE Transactions on Communications* 66 (3): 1139–1152.

12 Lei, H., Dai, Z., Ansari, I.S. et al. (2017). On secrecy performance of mixed RF-FSO systems. *IEEE Photonics Journal* 9 (4): 1–14.

13 Chan, V.W. (2003). Optical satellite networks. *Journal of Lightwave Technology* 21 (11): 2811.

14 ESOA 5G White Paper (2018). Satellite communications services: an integral part of the 5G ecosystem. https://www.esoa.net/cms-data/positions/5G%20infographic%20final_1.pdf (accessed 13 March 2021).

15 Pattanayak, D.R., Dwivedi, V.K., Karwal, V. et al. (2020). On the physical layer security of a decode and forward based mixed FSO/RF co-operative system. *IEEE Wireless Communications Letters* 9 (7): 1031–1035, https://doi.org/10.1109/LWC.2020.2979442.

16 Guidolin, F., Nekovee, M., Badia, L., and Zorzi, M. (2015). A study on the coexistence of fixed satellite service and cellular networks in a mmWave scenario. *2015 IEEE International Conference on Communications (ICC)*, pp. 2444–2449.

17 De Sanctis, M., Cianca, E., Araniti, G. et al. (2016). Satellite communications supporting internet of remote things. *IEEE Internet of Things Journal* 3 (1): 113–123.

18 Henry, C. (2018). Telesat says ideal LEO constellation is 292 satellites, but could be 512. https://spacenews.com/telesat-says-ideal-leo-constellation-is-292-satellites-but-could-be-512/ (accessed 13 March 2020).

19 Kusza, K.L. and Paluszek, M.A. (2000). Intersatellite Links: Lower Layer Protocols for Autonomous Constellations. Princeton Satellite Systems NJ, Tech. Rep. ADA451768.

20 del Portillo, I., Cameron, B.G., and Crawley, E.F. (2019). A technical comparison of three low earth orbit satellite constellation systems to provide global broadband. *Acta Astronautica* 159: 123–135.

21 Burleigh, S., Hooke, A., Torgerson, L. et al. (2003). Delay-tolerant networking: an approach to interplanetary Internet. *IEEE Communications Magazine* 41 (6): 128–136.

22 Yang, L., Gao, X., and Alouini, M. (2014). Performance analysis of relay-assisted all-optical FSO networks over strong atmospheric turbulence channels with pointing errors. *Journal of Lightwave Technology* 32 (23): 4613–4620.

23 Yang, L., Hasna, M.O., and Ansari, I.S. (2017). Unified performance analysis for multiuser mixed η - μ and $\mathcal{M}$ - distribution dual-hop RF/FSO systems. *IEEE Transactions on Communications* 65 (8): 3601–3613.

24 Boroson, D.M., Robinson, B.S., Murphy, D.V. et al. (2014). Overview and results of the lunar laser communication demonstration. *Free-Space Laser Communication and Atmospheric Propagation XXVI*, vol. 8971, International Society for Optics and Photonics, p. 89710S.

25 Jain, S., Fall, K., and Patra, R. (2004). Routing in a delay tolerant network. *SIGCOMM Computer Communication Review* 34 (4): 145–158. https://doi.org/10.1145/1030194.1015484.

26 Cruz-Sánchez, H., Franck, L., and Beylot, A.-L. (2010). Routing metrics for store and forward satellite constellations. *IET Communications* 4 (13): 1563–1572.

27 Jianwei, L., Weiran, L., Qianhong, W. et al. (2016). Survey on key security technologies for space information networks. *Journal of Communications and Information Networks* 1 (1): 72–85.

28 Arslan, M.G. and Alagoz, F. (2006). Security issues and performance study of key management techniques over satellite links. *2006 11th International Workshop on Computer-Aided Modeling, Analysis and Design of Communication Links and Networks*, pp. 122–128.

29 Howarth, M.P., Iyengar, S., Sun, Z., and Cruickshank, H. (2004). Dynamics of key management in secure satellite multicast. *IEEE Journal on Selected Areas in Communications* 22 (2): 308–319.

30 Barker, E., Smid, M., Branstad, D., and Chokhani, S. (2013). A framework for designing cryptographic key management systems. *NIST Special Publication* 800 (130): 1–112.

31 Qi, X., Ma, J., Wu, D. et al. (2016). A survey of routing techniques for satellite networks. *Journal of Communications and Information Networks* 1 (4): 66–85.

32 Alagoz, F., Korcak, O., and Jamalipour, A. (2007). Exploring the routing strategies in next-generation satellite networks. *IEEE Wireless Communications* 14 (3): 79–88.

33 Usman, M., Qaraqe, M., Asghar, M.R., and Ansari, I.S. (2020). Mitigating distributed denial of service attacks in satellite networks. *Transactions on Emerging Telecommunications Technologies* 31 (6): e3936. https://onlinelibrary.wiley.com/doi/abs/10.1002/ett.3936.

34 Onen, M. and Molva, R. (2004). Denial of service prevention in satellite networks. *2004 IEEE International Conference on Communications (IEEE Cat. No. 04CH37577)*, Volume 7, IEEE, pp. 4387–4391.

35 Zargar, S.T., Joshi, J., and Tipper, D. (2013). A survey of defense mechanisms against distributed denial of service (DDoS) flooding attacks. *IEEE Communications Surveys & Tutorials* 15 (4): 2046–2069.

36 Yang, L., Hasna, M.O., and Ansari, I.S. (2018). Physical layer security for TAS/MRC systems with and without co-channel interference over η-μ fading channels. *IEEE Transactions on Vehicular Technology* 67 (12): 12421–12426.

37 Lei, H., Luo, H., Park, K. et al. (2020). On secure mixed RF-FSO systems with TAS and imperfect CSI. *IEEE Transactions on Communications* 68 (7): 4461–4475.

38 Yang, L., Hasna, M.O., and Gao, X. (2015). Performance of mixed RF/FSO with variable gain over generalized atmospheric turbulence channels. *IEEE Journal on Selected Areas in Communications* 33 (9): 1913–1924.

39 Oi, D.K., Ling, A., Vallone, G. et al. (2017). Cubesat quantum communications mission. *EPJ Quantum Technology* 4 (1): 6.

40 Bacsardi, L. (2013). On the way to quantum-based satellite communication. *IEEE Communications Magazine* 51 (8): 50–55.

14

Conclusion

Airborne wireless networks that are enabled by unmanned aerial vehicles (UAVs) can provide cost-effective and reliable wireless communications to support the use cases in future networks. Despite the several benefits, airborne wireless networks suffer from some realistic constraints such as being energy constrained because of the limited battery power, safety concerns, and the strict flight zone. In this book, we have curated a list of chapters that touch upon the propagation channel for the airborne networks, self-organized UAVs, and some of its use cases including disaster recovery, agriculture, underwater monitoring, and smart lockdown monitoring. We briefly discuss some of the hot research topics that have surfaced.

14.1 Future Hot Topics

14.1.1 Terahertz Communications

It is now well established that the sub-3 GHz spectrum bands used in legacy wireless communications systems are insufficient to meet the high data rate requirement of emerging mobile applications. Terahertz communication is a promising candidate for meeting the high data rate requirement in the next-generation communication systems. Its application is limited to offering ultra-broadband at short range; hence the performance of UAVs and drones in the Terahertz networks deserves attention. The coverage probability, area spectral efficiency, and the line-of-sight probability model need to be analyzed for the airborne Terahertz networks since a severe degradation in path loss is expected. Further, the impact of the UAV density on the coverage and area spectral efficiency also should be researched.

Autonomous Airborne Wireless Networks, First Edition.
Edited by Muhammad Ali Imran, Oluwakayode Onireti, Shuja Ansari, and Qammer H. Abbasi.
© 2021 John Wiley & Sons Ltd. Published 2021 by John Wiley & Sons Ltd.

14.1.2   3D MIMO for Airborne Networks

The UAV network can be seen as an airborne antenna that can be exploited for technologies such as millimeter-wave communications, massive multiple-input multiple-output (MIMO), and 3D MIMO. The latter can yield higher system capacity and support a higher number of users, thus making it very suitable for airborne networks. The application of 3D MIMO on airborne wireless networks needs to be investigated not only in terms of spectral efficiency but also in terms of energy efficiency. The tradeoff between spectral efficiency and energy efficiency for the 3D MIMO airborne network deserves attention in a future study. Such a study could extend to both the 3D MIMO millimeter-wave and the Terahertz airborne networks. Furthermore, the impact of the altitude of the UAVs and the line of sight (LoS) channel conditions (between the UAVs and users on the ground) in enabling the effective beam in both the elevation and the azimuth plane/domain also deserves attention in a future study.

14.1.3   Cache-Enabled Airborne Networks

Caching at the small cell base station has been established as a technique for reducing delay and improving the throughput of the user. However, such a static caching approach has a significant limitation in dense networks with mobility due to handover and the need to cache the content in multiple base stations. Cache-enabled airborne network where the popular content can be dynamically cached at the UAV while also tracking the mobility pattern of the users can improve both the quality of service and the quality of experience. The gains can be further increased through a proactive positioning of cache-enabled UAVs to deliver the UE services.

14.1.4   Blockchain-Enabled Airborne Wireless Networks

UAVs and drones can be used to carry sensors and software, which can then fly over a defined area and capture the required data. In this use case, it is important to prevent malicious entities from altering the transmitted data in airborne network communications. The current systems in airborne networks are with various vulnerabilities and can be hacked by people with some set skills. Blockchain uses the combination of a consensus algorithm, distributed ledger, and cryptography to create a decentralized trustworthy platform. The security limitations and the challenges of the current blockchain networks when used for the airborne networks deserve attention in a future study.

14.2   Concluding Remarks

The use of airborne wireless networks enabled by UAVs is desirable due to their high maneuverability, ease of operation, and high affordability for various civilian applications, such as disaster relief, aerial photography, remote surveillance, and continuous telemetry. The airborne wireless networks can support the terrestrial cellular network's utility with its promising attributes. UAVs integrated with cellular wireless networks can maximize the data rate, and improve the coverage and network capacity. Thus, the hard and high-performance requirements of future heterogeneous cellular networks could be fulfilled by UAVs supporting different applications. This book enables the readers with the fundamental theories of the airborne wireless network, its propagation channel model, its applications in emergency and disaster recovery, underwater monitoring, and smart farming. Throughout the chapters, the book presents open issues and challenges, addressing which can further advance the field.

Index

Autonomous Airborne Wireless Networks, First Edition.
Edited by Muhammad Ali Imran, Oluwakayode Onireti, Shuja Ansari, and Qammer H. Abbasi.
© 2021 John Wiley & Sons Ltd. Published 2021 by John Wiley & Sons Ltd.